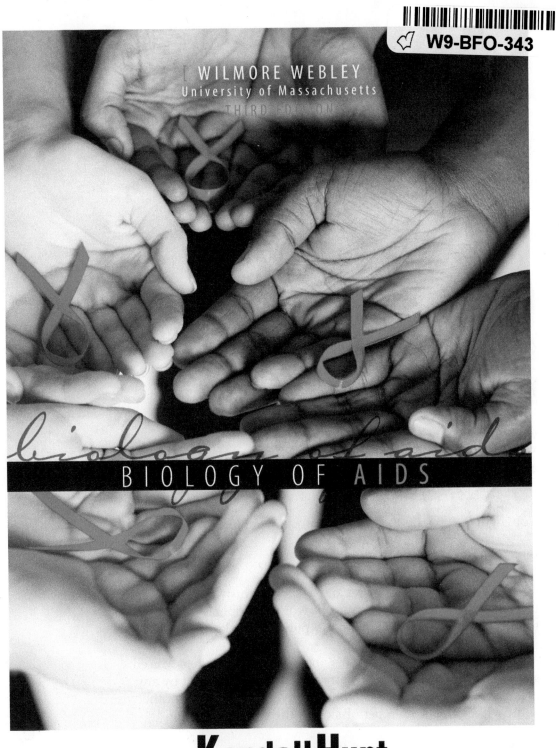

WILMORE WEBLEY
University of Massachusetts

THIRD EDITION

BIOLOGY OF AIDS

Kendall Hunt
publishing company

Kendall Hunt
publishing company

www.kendallhunt.com
Send all inquiries to:
4050 Westmark Drive
Dubuque, IA 52004-1840

Copyright © 2007, 2008, 2013 by Wilmore Webley

ISBN 978-1-4652-2838-3

Printed in the United States of America
10 9 8 7 6 5 4 3 2 1

I dedicate this book to:

CNM: Your goals, dreams and highest aspirations are still within reach and you are more ready for the challenges of living with HIV than you realize. You will always have my unconditional support.

And

My wonderful, funny, smart and caring son Zachary Webley. I pray that you will always be filled with compassion, overflow with kindness, and realize the value of every life as you interact with people each day of your existence.

Contents

CHAPTER 1

HIV/AIDS: The Early Years

In the Beginning: Discovery, Confusion, and Speculations

In the not-so-distant past, it was almost a guarantee that you would die of an infectious disease. In fact, even at the turn of the 1900s your chances of dying of an infectious disease before age 10 was extremely high. These facts reiterate that coping with the constant threat of infectious disease outbreaks has always been an integral part of the history of man. While many fear that asteroids or aliens from outer space might one day destroy our planet, the greatest threat to our existence as a species comes from ever-evolving microscopic organisms seeking to colonize our bodies in order to propagate their own species. However, this was not the prevailing sentiment in the 1960s to 1980s. The most challenging microbial pathogens of our time had fallen to the constant onslaught of the vaccine revolution as well as new and powerful antibiotics. Diseases such as smallpox, polio, mumps, measles, rubella, tetanus, and diphtheria that had wreaked havoc on many parts of the world were under control, and many believed that we would never have to contend with a microbial threat that we could not handle again. That was until a June 5th, 1981, Centers for Disease Control and Prevention (CDC) publication in the *Morbidity and Mortality Weekly Report* (*MMWR*) changed everything. The report described the unusual case studies of five young men with an unusual illness. During the period from October 1980 to May 1981, five young men, all active homosexuals, were treated for biopsy-confirmed *Pneumocystis carinii pneumonia* (PCP) at three different hospitals in Los Angeles, California. Two of the patients died. All five patients had laboratory-confirmed previous or current cytomegalovirus (CMV) infection and mucosal Candidiasis (yeast) infection. The authors of the *MMWR* report, led by Dr. Micheal Gottleib at the University of California at Los Angeles, indicated that all of these patients exhibited signs of severe immunodeficiency not normally seen in this age group. The *New York Times* published an article on July 3, 1981, entitled "Rare Cancer Seen in 41 Homosexuals" describing the unusual occurrence of a rare skin cancer, Kaposi's sarcoma, seen in homosexual men. A day later, on July 4, the CDC reported that during the preceding 30 months, 26 cases of Kaposi's sarcoma had been reported among gay males, and that eight had died, all within 24 months of diagnosis. These reported infections, both in the *MMWR* and the *New York Times*, were later determined to be some of the first cases of infection with the human immunodeficiency virus (HIV), which eventually leads to a state of dysfunction in the body's immune functions called acquired immunodeficiency syndrome (AIDS); thus began what would become the most deadly pandemic of the 20th century.

Since 1981 more than 65 million people worldwide have been infected with the virus, including the approximately 34 million people currently living with the disease (as of 2010). According to 2009 figures, approximately 30 million people have died of AIDS since 1981. While 1981 is generally referred to as the beginning of the HIV/AIDS epidemic, scientists now believe that HIV was present years before the first case was brought to public attention. Ronald Reagan, U.S. president at the time, refused to talk about the growing problem of AIDS. Nonetheless, in 1982 the first congressional hearings were held on HIV/AIDS. For a while the disease term "GRID," or "gay-related immune deficiency," was used by the media and healthcare professionals, as it was mistakenly thought that there was an inherent link between homosexuality and the syndrome, and only homosexuals were susceptible. The CDC convened a meeting of scientists, blood industry executives, gay activists, hemophiliacs, and others to develop guidelines for screening the blood supply. Although they decided then to adopt a "wait-and-see" attitude, they formally established the term *acquired immune deficiency syndrome (AIDS)* in 1982, referring at that time to four "identified risk factors" of male homosexuality, intravenous drug abuse, Haitian origin, and hemophilia A. Doctors agreed that the term AIDS was appropriate since people acquired the condition rather than inherited it; because it resulted in immune system deficiency; and because it was a syndrome, characterized by a number of manifestations, rather than a single disease. On April 22, 1984, Dr. James Mason of the CDC was reported as saying, "I believe we have the cause of AIDS." He was referring to a virus called lymphadenopathy virus (LAV) isolated in 1983 by Luc Montagnier of the Pasteur Institute in France. A day later, United States Health and Human Services Secretary Margaret Heckler announced that Dr. Robert Gallo (see Figure 1-1.) of the National Cancer Institute had isolated the virus that caused AIDS, a virus Gallo named human T-cell leukemia virus type 3 (HTLV-III).

Heckler stated that there would soon be a commercially available test able to detect the virus and that a vaccine to combat the virus would be available for testing within two years. Heckler concluded *"yet another terrible disease is about to yield to patience, persistence, and outright genius."* The virus was later named the human immunodeficiency virus (HIV). Still very little was known about modes of transmission, and public anxiety continued to grow. In the same year, the CDC issued a statement suggesting that abstention from intravenous drug use and reduction of needle-sharing should be effective in preventing transmission of the virus. This proved to be correct as time passed. It was not until September 17, 1985, that then-president Ronald Reagan mentioned the word "AIDS" in

Figure 1-1 Robert Gallo, co-discoverer of the HIV virus, in the early 1980s among (from left to right) Sandra Eva, Sandra Colombini, and Ersell Richardson.

public for the first time in response to a reporter's query. In this same year, the Food and Drug Administration (FDA) approved the first HIV antibody test, and routine tests on blood products in the United States and Japan commenced. It was also in 1985 that the first International Conference on AIDS was held in Atlanta, Georgia, and thus commenced the intense collaborations between and among several agencies of the U.S. government (www.avert.org/his8186.htm).

Current State of the AIDS Epidemic

Today, there are over 1.2 million persons living with HIV in the United States. About one-fourth of those with HIV have not yet been diagnosed and are unaware that they are infected. Since those very early beginnings, we have learned much about HIV, the infectious cause of AIDS, as well as about AIDS, the disease. Consequently, HIV has become the most widely studied virus (see Figure 1-2.), and as a direct result of the extensive genetic and proteomic analyses performed on the virus, we currently have over 31 FDA-approved drugs that are used to treat HIV infections in the United States.

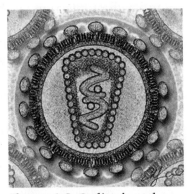

Figure 1-2 Stylized rendering of a cross section of the AIDS virus.
Source: http://en.wikipedia.org/wiki/HIV.

Unlike the disorganized panic that ensued in the early 1980s, current investigators understand a great deal about the spread of the virus in our population.

Most notably, it is now clear that HIV transmission requires close contact with an infected person and that infection occurs most frequently through the exchange of body fluids. This has led many to speculate that AIDS can now be viewed as a chronic disease, just like diabetes, that is quite manageable. While this might be partially true in some developed parts of the world where antiretroviral drugs are readily available and patients can afford them, a quick glance at the worldwide incidence, prevalence, and mortality statistics indicate that we are still faced with a monumental global health crisis. According to the latest data from the Joint United Nations Program on HIV/AIDS (UNAIDS) and the World Health Organization (WHO), as of the end of 2011, there were 34 million [31.4 million–35.9 million] people living with HIV/AIDS worldwide. It has been estimated that 0.8% of adults aged 15-49 years worldwide are living with HIV. Although the number of people living with the disease continues to increase, the current estimates indicate that the global HIV/AIDS prevalence rate (the percent of people living with the disease) has leveled off. The fact is that the burden of the epidemic continues to vary considerably between countries and regions. Sub-Saharan Africa remains the most severely affected region, accounting for 69% of the people living with HIV worldwide. At the same time, almost 5 million people are living with HIV in the combined regions representing South, Southeast and East Asia. The Caribbean, Eastern Europe, and Central Asia are a cause for concern with 1.0% of adults living with HIV at the end of 2011. Worldwide, 2.5 million [2.2 million–2.8 million] people became newly infected with HIV in 2011 and approximately 1.7 million [1.5 million– 1.9 million] people died from AIDS-related causes worldwide—24% fewer deaths than in 2005. This most welcome decline is reflective of the overall reduction in the total number of new cases of the disease. In 2011 the number of newly infected individuals was 20% lower than in 2001. There has been a 42% reduction in new HIV infections in the Caribbean, which has the second highest infection rate behind sub-Saharan Africa. It is also very encouraging to see that half of all reductions in new HIV infections over the last two years have been among newborn children. This demonstrates that that elimination of new infections in children is indeed possible. In 2011, new infections in children were 43% lower than in 2003. This decline is however not uniform across the world. Since 2001, the number of people newly infected in the Middle East and North Africa has increased by more than 35%.

Just about half of the adults living with HIV/AIDS worldwide are women, while young people under the age of 25 are estimated to account for half of all new HIV infections worldwide. It is estimated that every day over 6,800 persons become newly infected with HIV and over 5,700 die from AIDS. Notably, nearly 69% of the people with HIV/AIDS live in sub-Saharan Africa, although this region has only 10% of the world's population. This region is also home to almost 95% of the world's AIDS orphans. Women account for 58% of people living with HIV in sub-Saharan Africa. For most of the AIDS patients in sub-Saharan Africa and Southeast Asia, there is no adequate healthcare system and antiviral drugs are either not available or are too expensive for the average person. AIDS has killed more than 30 million people since it was first recognized in 1981, making it one of the most destructive epidemics in recorded history (Statistics from UNAIDS). HIV/AIDS is the leading cause of death worldwide for people aged 15–59, and half of all new infections occur in people under the age of 25 (www.pbs.org/wgbh/ pages/ frontline/aids/etc/synopsis.html). There are also key populations within any country that continue to be vulnerable to high incidence and prevalence of HIV infection and AIDS. Around 3 million of the estimated 16 million people who use drugs are living with HIV. Analysis of data acquired from 49 countries confirm that people who inject drugs are among the population groups most severely affected by HIV infection. In virtually all countries reporting data in 2012, the prevalence of HIV infection is higher among people who inject drugs than among the general population. HIV prevalence was 22 times higher in people who use drugs than in the general population. In Eastern Europe and central Asia, domestic public sector sources provide only 15% of spending on prevention programs for people who inject drugs, although injection drug use has been a major source of new HIV infections in these regions over the past decade.

In areas where HIV is considered a generalized epidemic (meaning that more than 1% of the population is HIV positive), prevalence is consistently higher among sex workers in the capital cities than in the general population, and hovers at around 23%. Other vulnerable populations include men who have sex with men living in large cities, where the HIV prevalence is on average 13.5 times higher than in the general population (current figures are from unaids.org). Men who have sex with men (MSM) continue to represent a significant HIV risk group. The latest statistics continue to show that one of the major reasons for the persistent epidemic among MSM is that levels of consistent condom use are insufficient. The median proportion of MSM receiving an HIV test in the past 12 months is 38%.

Governments, healthcare officials, public and private funding agencies, and partners worldwide have embarked on a historic effort to end new HIV infections among children and reduce the number of women living with HIV who die from pregnancy-related causes. In the three years 2009 to 2011, antiretroviral prophylaxis prevented 409,000 children from acquiring HIV infection in low- and middle-income countries. However, much more can and need to be done. Only 30% of eligible pregnant women were receiving antiretroviral therapy for their own health in 2011, compared with 54% for all eligible adults. Children of mothers in marginalized populations experience HIV transmission rates nearly 2.5 times higher than in the general population.

In 2011, for the first time, a majority (54%) of people eligible for antiretroviral therapy in low- and middle-income countries were receiving it. HIV treatment coverage is 68% for women and 47% for men in low-and middle-income countries, compared with 28% for children worldwide. Major strides have been made toward the global goal of reducing the number of tuberculosis (TB)-related deaths among people living with HIV by 50% by 2015. The number of TB-related deaths among people living with HIV has fallen by 25% in the last decade. However, globally in 2011, fewer than half (48%) of the people with TB with a documented HIV-positive test result obtained antiretroviral therapy.

HIV continues to be one of the most deadly viruses humankind has ever faced. But, it has one weakness: Infection by the virus is absolutely preventable by relatively basic lifestyle measures. These include using condoms and clean needles, proper testing and protection of the blood supply, and testing and administering of antiretroviral medicine to expectant mothers. After more than three decades of research and drug development, the inescapable truth is that the only way to win the battle with HIV and AIDS is to reduce the number of people becoming infected in the first place. Currently, prevention is still the only "cure" available for HIV/AIDS. Many experts and government officials therefore agree that a comprehensive approach must be adopted to prevent further spread of HIV and AIDS. HIV prevention strategies must include close monitoring of the epidemic to target prevention and effective patient care activities. There needs to be better research on the effectiveness of current and evolving prevention methods. At the same time there has to be a better system for disseminating proven effective interventions, funding the implementation and evaluation of prevention efforts in high-risk communities, encouraging early diagnosis of HIV infection, and fostering meaningful collaborations between prevention and treatment programs. Although 95% of all infections take place in the developing world with less organized and funded healthcare systems, HIV today is a threat to men, women, and children in every corner of the world. For this reason, World AIDS Day (see Figure 1-3 for World

Figure 1-3 World AIDS Day is December 1st of each year.

AIDS Day logo), which started on the first of December 1988, is not just a fundraising event, but a deliberate attempt to increase awareness, fight prejudice, and improve HIV/AIDS education. This day is necessary to remind people that HIV has not gone away, and that there is much to be done yet. If these measures are not taken, while we yet sleep, this epidemic will creep upon us like a thief in the night, substantially altering the life goals and aspirations of the greatest hope for the future of our world; the youth in our communities.

Test Your Knowledge

Name: _____

1. True or False: Coping with the constant threat of infectious disease outbreaks has always been an integral part of the human history.

2. True or False: The first International Conference on AIDS was held in Atlanta, Georgia in 1985.

3. True or False: HIV is well controlled globally, therefore AIDS can now be viewed as a chronic disease, just like diabetes, that is quite manageable.

4. What is the difference between HIV and AIDS?
   (A) AIDS is the final stage of HIV infection.
   (B) There is no difference between HIV and AIDS.
   (C) HIV is an opportunistic virus seen in people with AIDS.
   (D) HIV is the virus that causes AIDS.
   (E) Both A and D.

5. The first AIDS patients encountered in San Francisco and New York presented with _____.
   (A) shortness of breath
   (B) high HIV viral load
   (C) Kaposi's sarcoma
   (D) low CD4 cell counts
   (E) both B and C

6. Sub-Saharan Africa accounts for what percentage of all HIV infections?
   (A) 25%
   (B) 58%
   (C) 69%
   (D) 72%
   (E) 88%

7. According to UNAIDS, approximately how many people were living with HIV/AIDS worldwide at the end of 2011?

8. In what year did the CDC report on a strange new disease that later became known as AIDS?

CHAPTER 2

Understanding HIV and AIDS

What Is AIDS?

AIDS stands for acquired immunodeficiency syndrome. *Acquired*, which means you catch it; *immunodeficiency* means that your immune system is weakened and can no longer effectively defend the body against infectious agents; and *syndrome*, describing a combination of health problems that collectively point to a particular disorder. AIDS is caused by the human immunodeficiency virus (HIV). By infecting and eventually leading to the destruction and/or functional impairment of cells of the immune system—notably helper T cells—HIV progressively destroys the body's ability to fight common infections and certain cancers caused by infectious agents. If left untreated, AIDS leads to death. Nobody dies directly from infection with HIV itself, but from the illnesses that develop due to the destruction of the immune system by the virus. The term *AIDS* therefore refers to an advanced stage of HIV infection, when the immune system has sustained substantial damage, thereby leaving the body vulnerable to opportunistic infections. This means, therefore, that not everyone with HIV has AIDS. We now know that it takes approximately 10 years on average from the time of initial infection with HIV for the onset of the symptoms of AIDS to appear. Moreover, with consistent use of current antiretroviral drugs and close monitoring of specific immune parameters, it is expected that many individuals who are HIV positive will live their entire lives without progressing to AIDS.

What Are Viruses?

When the word *virus* is used today, most people automatically think about their computers and terms like McAfee and Norton antivirus software start popping into our heads. For those born before the age of computer viruses, they might think of virus pandemics caused by smallpox, rubella, and influenza. In 1884, Charles Chamberland, in the lab of Louis Pasteur, discovered that if you passed liquid containing bacteria through an unglazed porcelain tube, the bacteria would be retained and the solution that passed through (the filtrate) was sterile. This technique became a major form of sterilization until 1892 when Dmitri Iwanowski applied this test to a filtrate of plants suffering from Tobacco Mosaic Disease with surprising results. The filtrate was capable of producing the original disease in new plants! Repeated filtrations produced the same results and nothing could be seen in the filtrates using the most powerful microscopes at that time. They also realized that they could not culture anything from the filtrates using nutrient agar in Petri plates as they were accustomed to doing with filtered bacteria. Iwanowski and his colleagues subsequently concluded that they had discovered a new pathogenic life-form, which they called "Filterable Virus." A virus in this context is an ultramicroscopic (very small), biological infectious agent that lacks the means for self-reproduction outside a host cell. Viruses are therefore called obligate intracellular, but unlike other obligate intracellular parasites (such as *Chlamydia*), viruses are not truly alive, although some may disagree. Although many viruses are pathogenic (cause disease), most viruses in nature do not cause disease in humans. There are viruses that infect animals, plants, insects, and even bacteria (bacteriophages). You can go to http://en.wikipedia.org/wiki/Listofviruses for a list of known viruses and the diseases they cause. The truth is that a virus does not mean to cause harm to its host. It is just interested in making as many copies of itself as possible. It is programmed to do this in order to maintain that species in the population. When a virus kills its host, unless it has been caught by a second living host, the virus will also eventually die, defeating its true purpose of mass replication in the first place. A successful pathogen never kills its host before it can exit and enter another susceptible host.

At the most basic level, a virus consists of a piece of nucleic acid material (DNA or RNA; never both!), wrapped in a thin coat of protein called the viral capsid. The nucleic acid material can either be double- or single-stranded. Many, but not all, viruses also have an outer coat called an envelope derived from the infected cell membrane. They range in size from 20 to 250 nanometers in diameter (one nanometer is one-billionth of a meter). Compare this size to a typical bacterial cell (prokaryotic cell), which measures 1 to 10 micrometers, or the average animal cell (eukaryotic cell), which is approximately 10 to 100 micrometers in diameter (1 micrometer = 1,000 nanometers). See Figure 2-1. Outside of a living cell, a virus is dormant, but once inside, it splices its nucleic acid material into the host cell's DNA and with access to the resources of the host cell, initiates the production of more virus particles. To enter

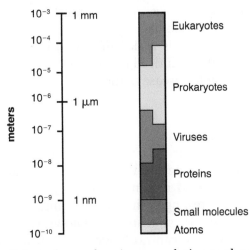

Figure 2-1 The range of sizes shown by viruses, relative to those of other organisms and biomolecules.
(Source: http://en.wikipedia.org/wiki/Image:Relative_scale.svg)

a host cell, a viral particle first attaches to a receptor protein on the host cell membrane using an attachment protein that is essential to its survival.

If a virus cannot attach to the cell, it cannot enter and therefore cannot cause disease. These proteins on the host cell to which viral particles attach were not created for viral entry and many times serve very important cellular functions. The viruses have just evolved to use already existing receptors to enter the host cell. For this reason, there are microbes, including viruses that are able to infect certain cells and not others or that can cause disease in one animal but not others. Humans do not get canine distemper, caused by a RNA virus that is closely related to the human measles virus and can range from mild cold-like symptoms in dogs to severe neurological signs including seizures. Humans, however, are not susceptible to this disease because we do not have the cellular proteins required for rapid viral replication or have developed antibodies and other cellular immune responses to it as a result of the measles vaccine that we routinely get.

Following attachment, the virus fuses with the cell membrane and extrudes its nucleic acid material into the cell where it integrates into the cellular genome. The virus then utilizes cellular transcription, translation, as well as energy machinery to replicate its genome and produce new viral proteins. The newly formed viral components assemble to form new virions or viruses. Some viruses exit the cells by budding through the cell membrane, thereby acquiring a lipid bilayer in the process. Other viruses, however, exit the cell by releasing enzymes that cause cell lysis or rupture and do not acquire a lipid bilayer. The viruses that bud from the cell membrane are called enveloped viruses and those that cause host cell rupture to exit the cell are said to be non-enveloped or naked viruses. A single infected host cell may produce up to 10,000 new viral particles. As previously stated, a virus therefore does not really intend to cause harm; all it wants to do is make more viruses. Unfortunately for us, in its attempt to preserve itself, harm is indeed caused to the individual infected, as well as uninfected host cells, leading to disease pathologies.

The Nature of HIV

HIV hails from the *Lentivirus* family of viruses, all of which are retroviruses and share distinct characteristics, including the ability to systematically attack cells of the immune system. The viruses can be found in a number of different animals, including cats, sheep, horses, and cattle. Interestingly, the simian immunodeficiency virus (SIV) that infects monkeys is a lentivirus. These viruses are known to steadily weaken the body's defense (immune) system until it can no longer fight off infections such as pneumonia, diarrhea, tumors, and other illnesses, all of which can be part of the AIDS (acquired immunodeficiency syndrome) complex.

An HIV particle is approximately 100–150 billionths of a meter in diameter. That is approximately 0.1 microns or 4 millionths of an inch! This means that HIV is one-seventieth of the diameter of the human CD4+ white blood cells that they typically infect (see TEM image in Figure 2-2.). These cells are called CD4 because of a protein that is ubiquitously expressed on the cell surface of T-helper cells. HIV particles surround themselves with a viral envelope through which project approximately 72 spikes that are formed from the gp120 and gp41 proteins (Figure 2-3).

The "gp" in gp120 or gp41 refers to the glycoprotein nature of the molecules and infers that they are a combination of carbohydrate and protein molecules joined together by chemical linkage. Immediately below this outer viral envelope is

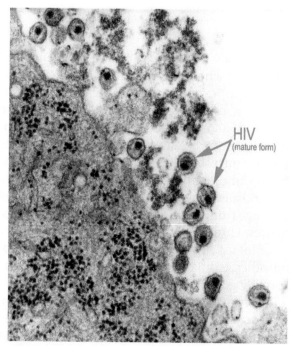

HIV
(mature form)

Figure 2-2 This highly magnified transmission electron micrographic (TEM) image reveals the presence of mature forms of the human immunodeficiency virus (HIV) in a tissue sample under investigation.
(*Source: CDC Public Health Image Library http://phil.cdc.gov/phil/home.asp.*)

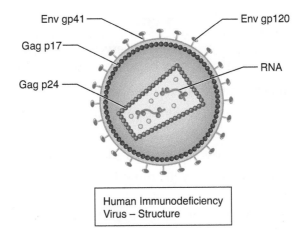

Env gp41

Env gp120

Gag p17

RNA

Gag p24

Human Immunodeficiency
Virus – Structure

Figure 2-3 This illustration shows the structure of the human immunodeficiency virus (HIV). The outer shell of the virus is known as the viral envelope. Embedded in the viral envelope is a complex protein known as env, which consists of an outer protruding cap glycoprotein (gp) 120, and a stem gp41. Within the viral envelope is an HIV protein called p17 (matrix), and within this is the viral core or capsid, which is made of another viral protein p24 (core antigen). The major elements contained within the viral core are two single strands of HIV RNA, a protein p7 (nucleocapsid), and three enzyme proteins, p51 (reverse transcriptase), p11 (protease), and p32 (integrase). *(Source: Courtesy of AVERT.org. www.avert. org/virus.htm.)*

the viral matrix, which is made from the protein p17. The viral core (or capsid) is usually bullet-shaped and is made from the p24 protein. Inside the core are three enzymes required for HIV replication called reverse transcriptase, integrase, and protease. Also held within the core is HIV's genetic material. While most organisms, including viruses, store their genetic material on strands of DNA, retroviruses like HIV are the exception because their genes are composed of RNA (ribonucleic acid). The HIV genome consists of two identical strands of RNA, making the HIV replication process more complicated than that seen in most other viruses.

HIV is on the smaller end of the scale where viruses are concerned. Compared to bacteria that generally have over a thousand genes and a human cell containing approximately 30,000 genes, HIV has just nine genes. Three of these HIV genes—*gag, pol,* and *env*—contain the information needed to make structural proteins (i.e., envelope glycoproteins, capsid, and polymerase, respectively) for new virus particles. The remaining six HIV genes, known as *tat, rev, nef, vif, vpr,* and *vpu,* code for proteins that control the ability of HIV to infect a cell, produce new copies of virus, or cause disease. The complete structure of an HIV-1 genome, extracted from infectious virions, has been solved to single-nucleotide resolution level, allowing scientists to determine any variations in clinical strains.

HIV Infection and Replication

The fusion, entry, and replication of HIV is very typical of retroviruses. The steps involved in the HIV replication cycle may be summarized as follows:

1. The HIV particle attaches and fuses to the host cell surface.
2. HIV RNA, reverse transcriptase, integrase, and other viral proteins enter.
3. Viral DNA is formed by reverse transcription.
4. Viral DNA is transported across the nucleus and integrated into the host DNA.
5. New viral RNA, which is used as genomic RNA and to make viral proteins, is produced.
6. Newly formed viral RNA and proteins are packaged and moved to cell surface and a new, immature, HIV virus forms.
7. The HIV protease cleaves viral proteins, resulting in maturation and release of individual HIV proteins.

Since HIV can only replicate inside human cells, the initial step in its infection involves attachment to a susceptible host cell. In fact, for any pathogen to cause disease, it has to first evolve a mechanism by which to attach to the host cells it intends to infect. Once attachment is accomplished, the pathogen may gain entry into the cell by various pathways. Although CD4+ helper T cells appear to be the main targets of HIV, other immune system cells are also infected. Cells of the mononuclear phagocyte system, principally blood monocytes and tissue macrophages, regulatory T lymphocytes, natural killer (NK) lymphocytes, dendritic cells (Langerhans cells of epithelia and follicular dendritic cells in lymph nodes), hematopoietic stem cells, endothelial cells, microglial cells in the brain, and gastrointestinal epithelial cells are all targets of HIV infection. Among these, the long-lived monocytes and macrophages have been shown to harbor large quantities of the virus without being killed. These cells therefore function as reservoirs of HIV. CD4+ T cells also serve as important reservoirs of HIV; a small proportion of these cells harbor HIV in a stable, inactive form. Normal immune processes may activate infected cells, resulting in the production of new HIV virions. This attachment event involves the specific interaction between the gp120 spike on HIV (see Figure 2-3) and CD4, an essential protein found on the surface of certain human immune cells.

What typically happens is that a viral particle gains entry to the body through infected body fluids, where it soon bumps into a cell that carries on its surface the CD4 protein. The gp120 spikes on the surface of the viral particle bind to the CD4 molecules, allowing the viral envelope to fuse with the host cell membrane. In addition, the viral protein then binds to a second kind of cell-surface molecule known as chemokine receptors. The HIV particle must interact with this second molecule on the cell surface to gain entry and commence a productive infection cycle. This second molecule is either CXCR4 or CCR5, depending on the cell type. The macrophage-tropic (M-tropic) HIV-1 strains utilize the chemokine receptor CCR5 in combination with

CD4 for entering target cells, while the T-tropic HIV-1 uses CXCR4. These molecules are therefore referred to as coreceptors for HIV. In fact, individuals in the population who are missing the gene for CCR5-(Delta32), do not produce a functional CCR5 protein and exhibit a near-complete protection against HIV-1 infection. This was the basis for the first functionally cured adult with HIV infection (see Chapter 10).

Once the virus attaches to its receptor and co-receptor, it sheds its envelope, and the nucleic acid material (RNA) is released into the cell where it is converted by the enzyme reverse transcriptase (RT). The process is called reverse transcription because the process of transcribing a protein typically goes from DNA to RNA and then to protein. However, since HIV is an RNA virus and needs to integrate its genes into the host cell DNA, this enzyme transcribes the single-stranded RNA into double-stranded DNA in the cytoplasm of the cell. The viral DNA then migrates into the nucleus, where it gets integrated into the host DNA by another HIV enzyme called integrase. HIV DNA that is spliced into the host cellular DNA is called a provirus. It is the job of the provirus to produce new viruses, which will be released from the infected cells. For this to happen, RNA copies must be made that can be read by the host cell's protein-making machinery. These copies are called messenger RNA (mRNA), and production of mRNA is called *transcription*, a process that involves the host cell enzymes. Viral genes in conjunction with the host cell machinery control this process. Viral genomic RNA is also transcribed for later incorporation into the newly assembled virus particle (Figure 2-4).

Following the transcription of viral mRNA in the cell's nucleus, this mRNA is transported to the cytoplasm where protein synthesis will take place. The HIV protein encoded by the *rev* gene is important here since it allows mRNA encoding the HIV structural proteins to migrate from the nucleus to the cytoplasm. Without the rev protein, structural proteins are not made. In the cytoplasm, the virus co-opts the cell's protein-making machinery to make long chains of viral proteins and enzymes, using HIV mRNA as a template in a process called *translation*. Newly synthesized HIV core proteins along with enzymes and genomic RNA are assembled inside the cell forming an immature viral particle. This immature virus then buds off from the cell, acquiring an envelope in the process that includes both cellular and viral proteins. During this part of the viral life cycle, the core of the virus is immature and the virus is not yet infectious. The long chains of proteins and enzymes that make up the immature viral core are now cut into smaller pieces by a viral enzyme called protease. This step results in infectious viral particles. One of the reasons HIV is so dangerous is that once it enters the CD4 cell, it hijacks the cell's machinery and turns out over 10 billion to 1 trillion new virus particles per day.

One of the defining properties of retroviruses is their ability to assemble into particles that can leave infected cells and spread to other susceptible cells and hosts. Viral budding from the host cell is a complex process. The morphogenesis of HIV can be divided into three stages. The first stage commences right after viral RNA and proteins are produced inside the cell and is referred to as the *assembly stage*, where the virion (virus particle, consisting of a capsid and an inner core of nucleic acid, but is

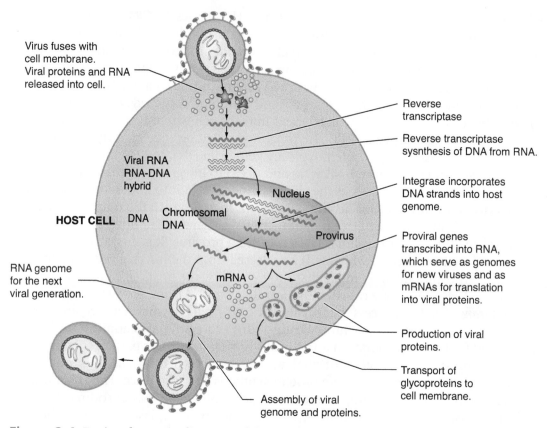

Virus fuses with
cell membrane.
Viral proteins and RNA
released into cell.

Reverse
transcriptase

Reverse transcriptase
sysnthesis of DNA from RNA.

Viral RNA
RNA-DNA
hybrid

Nucleus

Integrase incorporates
DNA strands into host
genome.

HOST CELL   DNA   Chromosomal
DNA

Provirus

Proviral genes
transcribed into RNA,
which serve as genomes
for new viruses and as
mRNAs for translation
into viral proteins.

RNA genome
for the next
viral generation.

mRNA

Production of viral
proteins.

Transport of
glycoproteins to
cell membrane.

Assembly of viral
genome and proteins.

Figure 2-4 Basic schematic diagram of the HIV replication cycle.

not yet infectious) is created and essential components are packaged. The next stage is *budding*, where the virion fuses with and crosses the plasma membrane obtaining its lipid envelope in the process. The final stage is *maturation*, wherein the virion becomes infectious through the aid of the **viral protease**. HIV protease cuts the newly synthesized **polyproteins** at appropriate sites to create the mature, individual, functional protein components of an infectious HIV **virion**. Without this protease, HIV virions would never become infectious and no new cells would be infected. As you might therefore imagine, the viral protease is a major target of HIV antiretroviral therapy. It is now believed that all of these stages are coordinated by the Gag polyprotein and its proteolytic maturation products, which function as the major structural proteins of the virus. Binding between the HIV Gag protein and molecules in the cell directs the accumulation of HIV components in special intracellular sacks, called multivesicular bodies (MVB), which normally function to carry proteins out of the cell. In this way, HIV actively hitchhikes out of the cell in the MVB by hijacking normal cell machinery. The good news is that discovery and better understanding of this budding pathway has revealed several potential points for intervening in the viral replication cycle.

Test Your Knowledge

Name: _____

1. Which of the following is true concerning the genetic material of a virus?
   (A) Only DNA viruses have nucleic acid.
   (B) It contains either DNA or RNA.
   (C) The genetic material of viruses is made of complex carbohydrates.
   (D) It contains both DNA and RNA.
   (E) Only RNA viruses cause disease.

2. What is the role of the HIV viral protease?
   (A) It must cut the host DNA to splice in viral DNA.
   (B) It stabilizes the newly made viral DNA.
   (C) It is needed by the reverse transcriptase.
   (D) It aids in attachment to the host cell membrane.
   (E) It cuts viral polyprotein into individual functional proteins.

3. Which statement about HIV/AIDS is true?
   (A) HIV is the same thing as AIDS.
   (B) The HIV organism lives inside cells and body fluids.
   (C) HIV organisms survive well in air and on surfaces.
   (D) Babies never contract HIV from their mothers.
   (E) Both A and B.

4. The Human Immunodeficiency Virus (HIV) _____.
   (A) is a DNA virus
   (B) has both DNA and RNA
   (C) is from the lentivirus family
   (D) is a bacteriophage
   (E) both A and C

5. In the process of HIV replication, Viral DNA is formed by _____.

6. True or False: AIDS patients never die directly from HIV pathology, but from the destruction of the immune system by the virus.

7. Who first discovered viruses?

8. What is the role of the HIV *gag, pol,* and *env* genes?

CHAPTER 3

The Origins of AIDS

From the very beginning, the origin of HIV has been the subject of intense debate and the cause of countless conspiracy theories. The first recognized cases of AIDS in the United States occurred in the early 1980s. These cases plunged the disease in the spotlight when several gay men in San Francisco and New York suddenly began to develop rare opportunistic infections and a rare cancer, Koposi's sarcoma, that were uncharacteristically resistant to conventional treatment regimens and not typically found in young, otherwise healthy individuals. Doctors soon discovered a distinctive feature of these cases. More than anything else, the men were lacking a specific type of white blood cell, which is essential to maintaining a healthy immune system. Healthy individuals typically have approximately 1,500 CD4 cells (also called T-helper cells) in each cubic milliliter of their blood. However, the men with this strange new disease typically had much lower levels, approximately 200 or less. This immune deficiency explained why they were so vulnerable to a plethora of diseases. It soon became clear that all the infected men were suffering from a common syndrome. The disease was formally called AIDS in 1982 and the discovery of HIV, the human immunodeficiency virus that causes AIDS, was made soon after. Some maintain that the virus was developed by the U.S. government to eliminate African Americans and gay people in New York City; others claim that the virus was genetically engineered to wage a biological warfare on Cuba, but then escaped and entered the general population; still others contend that HIV was developed through random mutations arising from secret electromagnetic warfare waged by the Soviet Union against the United States; while many still believe that the virus was spread either by a promiscuous flight attendant or through the oral polio vaccine program. So, what is the truth? Just where did HIV and its resulting disease, AIDS come from?

HIV/AIDS Dissidents

While most researchers today believe that the evidence supporting the fact that HIV causes AIDS is abundant and conclusive, there are many who disagree with this view. This has sparked many debates and has given rise to several groups of HIV/AIDS dissidents around the world. The AIDS dissident movement, also referred to as AIDS denialism, is a loosely connected group of individuals who dispute the scientific consensus that HIV is the cause of AIDS. Many dissident groups publicly deny the existence of HIV, while others accept that HIV exists, but argue that it is a harmless passenger virus and could therefore not be the cause of AIDS. Some dissidents further argue that the consensus that HIV causes AIDS has itself resulted in numerous inaccurate diagnoses leading to psychological terror and toxic treatments (referring here to AZT). They further charge that the commonly held view that HIV causes AIDS has led to a squandering of public funds, as well as an unprecedented deviation from the scientific method and standards. To date, the most vocal and probably the most significant scientist to question the HIV/AIDS theory is Professor Peter Duesberg (see Figure 3-1.) at the University of California at Berkeley.

Dr. Duesberg, a virologist, first published his opinion on this topic in 1988. Even with the large volume of data amassed by HIV/AIDS researchers throughout the 1990s to the present time, Dr. Duesberg remains unconvinced and is probably more defiant. While he concedes that HIV exists, he still maintains that it is harmless and is only present because the patient's immune system is already compromised by other factors. Dr. Duesberg believes that AIDS is caused mainly by drug abuse in developed countries of the world including the use of AZT, an antiviral drug currently used to combat HIV. For AIDS patients living in developing countries, like those on the African continent, Duesberg contends that malnutrition is the major factor at work here (www.duesberg.com). Dr. Duesberg is not alone; there are many more groups and individuals who believe as he does. Among his major supporters is Karry Mullis, a biochemist who won the 1993 Nobel Prize in recognition of his improvement of the polymerase chain reaction (PCR), a technique that allows a small strand of DNA to be copied almost an infinite number of times.

Figure 3-1 AIDS dissident Peter Duesberg has been a major force against the HIV/AIDS link.

© Roger Ressmeyer/CORBIS

Other prominent advocacy groups include the Perth Group led by Dr. Eleni Papadopulos (www.theperthgroup.com/), who do not believe that HIV really exists. The Perth group maintains that AIDS and all the associated HIV-induced phenomena are in fact caused by changes in cellular redox levels brought about by the oxidative nature of substances and exposures common to all the AIDS risk groups and to the cells used in the so-called "culture" and "isolation" of "HIV."

These dissident ideas have had significant and lasting impacts on the spread of the disease in South Africa. When the historic presidency of Nelson Mandela came to an end in 1999, he was succeeded by his then deputy, Thabo Mbeki. He at first showed strong support for the movement to eradicate HIV/AIDS from his people, even wearing an AIDS ribbon on his lapel the day he was sworn into office. However, by mid-1999 it was clear that his views had changed dramatically when he rose to give a speech in the upper house of South Africa's parliament. He started repeating statements apparently taken directly from Duesberg's writings, asking "How does a virus cause a syndrome? It can't . . . " In 2000, he invited several HIV/AIDS denialists, including Duesberg, to join his Presidential AIDS Advisory Panel. The scientific community at large was outraged by this move and responded with the Durban Declaration, a document affirming that HIV causes AIDS, signed by over 5,000 scientists and physicians. Still, Mbeki remained firm in his stance, refusing to allow access to medications to prevent mother-to-child transmission as well as timely treatment for those suffering from the disease. He had major support from health minister Manto Tshabalala Msimang and trade minister Alec Erwin. Today, South Africa still has one of the worst HIV/AIDS epidemics in the world. According to the UN, approximately 5.5 million people, or 18.8% of the adult population, are HIV positive. Independent studies have arrived at almost identical estimates of the human costs of HIV/AIDS denialism in South Africa. In a paper published in 2008, Pride Chigwedere and colleagues from the Harvard School of Public Health, concluded that between the years 2000 and 2005, more than 330,000 deaths and an estimated 35,000 infant HIV infections occurred "because of a failure to accept the use of available antiretroviral drugs to prevent and treat HIV/AIDS in a timely manner." This data was supported by others who got almost identical numbers, including Nicoli Nattrass of the University of Cape Town who estimates that 343,000 excess AIDS deaths and 171,000 infections resulted from the neglect of the Mbeki administration's policies. Mbeki was eventually forced to resign from the presidency on September 21, 2008, in large part because of his misguided policies on AIDS etiology. This is certainly a cautionary tale on information literacy and highlights the fact that ideas have consequences.

Despite the sometimes confusing message of these dissident groups, and the deadly toll of their influence in South Africa, there is more than enough evidence to support the fact that HIV infection indeed leads to the eventual development of AIDS. Several prominent scientists once associated with AIDS dissident groups have since changed their views and accepted the fact that HIV plays a role in causing AIDS, in response to an accumulation of new data. Probably most noteworthy of this group is Robert Root-Bernstein, author of *Rethinking AIDS: The Tragic Cost of Premature Consensus* and formerly a critic of the HIV/AIDS paradigm, has since distanced himself from the AIDS dissident movement, saying, "Both the camp that says HIV is a pussycat and the people who claim AIDS is all HIV are wrong . . . . The denialists make claims that are clearly inconsistent with existing studies. When I check the existing studies, I don't agree with the interpretation of the data, or, worse, I can't find the studies [at all]".

Evidence That HIV Causes AIDS

Among the numerous criteria used over the years to establish a causative link between a putative pathogenic (disease-causing) agent and a disease, the most cited are Koch's postulates, a set of rules for the assignment of a microbe as the cause of a disease. These rules were developed in the late 19th century by Robert Koch. Koch was a German physician who became famous for his discovery of the bacilli that cause anthrax, as well as the discovery of tuberculosis and cholera bacteria in the late 1800s. Robert Koch is considered one of the founders of bacteriology and was awarded the Nobel Prize in Physiology and Medicine for his tuberculosis findings in 1905. For more than a century, Koch's postulates have served as the litmus test for determining the cause of any epidemic disease. While they clearly have their limitations, the basic tenets of the postulates are as follows:

1. The suspected pathogen must always be found in diseased individuals, but absent in a healthy individual.
2. The pathogen must be isolated from the diseased individual and grown in a pure culture.
3. The pathogen from the pure cultures must cause the disease when inoculated into a healthy, susceptible host.
4. The same pathogen must be re-isolated from the host that was inoculated with the pure culture.

In order to address the question of whether HIV really causes AIDS, scientists have applied Koch's postulates as have been done throughout the past two centuries, although we admit that these postulates have inherent shortcomings. With regard to the first postulate, numerous studies from around the world show that HIV has been isolated from virtually every patient with AIDS. In addition, these patients are HIV-seropositive; meaning that they carry antibodies (highly specific immune defense proteins) against HIV, indicating that an HIV infection has taken place. These studies also demonstrated that the greater the HIV viral load (number of viral particles in the peripheral blood), the greater the likelihood of the patients to develop AIDS within a six-year period (Table 3-1).

With regard to postulate 2, which requires that the pathogen be isolated in pure culture, modern culture techniques have allowed the isolation of HIV from AIDS

Plasma RNA concentration (copies/mL of blood)	Proportion of patients who developed AIDS within 6 years
<500	5.4%
501–3,000	16.6%
3,001–10,000	31.7%
10,001–30,000	55.2%
>30,000	80.0%

Table 3-1 The greater the plasma RNA concentration indicating the presence of the virus, the greater the proportion of patients who progressed to AIDS within six years, indicating that there is a direct link between HIV viral load and onset of AIDS symptoms. Modified from Mallors et al, Ann Intern Med, 1997. **126**(12): pp. 946–54.

patients, as well as in HIV-seropositive individuals (those who have HIV-specific antibodies) with both early- and late-stage disease. In addition, the polymerase chain (PCR) and other molecular techniques that look for viral nucleic acid have enabled researchers to document the presence of HIV genes in patients with AIDS, as well as in individuals in earlier stages of HIV disease. While it is true that many patients in remote areas of the world die of AIDS without a diagnosis because of a lack of accessible healthcare, the evidence is overwhelming that HIV is the causative agent of AIDS.

Postulate 3 requires the isolated virus to be placed back into an uninfected host and observation of the same disease symptoms as seen in the previous host. This cannot be deliberately done by research scientists or medical doctors because of obvious ethical considerations. However, unfortunate accidental exposure to HIV in lab workers has resulted in the development of AIDS or severe immunosuppression after accidental exposure to concentrated, cloned HIV in the laboratory when no other risk factors were present.

In all cases, HIV was isolated from the infected individual, sequenced, and shown to be the same as the infecting strain of virus. In another tragic incident, transmission of HIV from a Florida dentist to six of his patients has been documented by genetic analyses of viruses isolated from both the dentist and the patients. The dentist and three of the patients developed AIDS and died, and at least one of the other patients has developed AIDS. Five of the patients had no HIV risk factors other than multiple visits to the dentist for invasive procedures. The development of AIDS following known HIV infection through seroconversion also has been repeatedly observed in pediatric as well as adult blood transfusion cases. Infections caused by mother-to-child transmission, accidental infection of hemophilia patients, injection-drug use, and sexual transmission in which seroconversion can be documented using serial blood samples, consistently document the fact that HIV isolated from infected individuals can lead to AIDS in previously uninfected individuals.

Finally, and maybe most importantly, no other agent or condition, including viral infections, bacterial infections, sexual behavior patterns, and drug abuse patterns predict who develops AIDS besides infection with HIV. Individuals from diverse backgrounds, including heterosexual men and women, homosexual men and women, hemophiliacs, sexual partners of hemophiliacs and transfusion recipients, injection-drug users, the elderly in our society, and infants have all developed AIDS, with the only common denominator being their prior infection with the agent that leads to HIV/AIDS (www.niaid.nih.gov/publications/hivaids/hivaids.htm). When the HIV/AIDS epidemic began in the early 1980s, people with HIV were not likely to live more than a few years. Today, there are over 31 antiretroviral drugs approved by the Food and Drug Administration (FDA) to treat HIV. While these drugs do not cure HIV infection, they have been shown to suppress the viral replication, decreasing the risk of transmission between sexual partners as well as mother-to-child transmission. Currently, there are HIV-infected individuals who have been living with the infection for over 30 years. The question then is, if HIV does not cause AIDS, why do anti-HIV drugs delay and or prevent the development of AIDS? With this accumulated evidence clearly showing that AIDS does not occur in the absence of HIV, that HIV infection is the only universal factor that predicts who will develop AIDS, coupled with surveillance statistics and the tremendous success of modern antiretroviral therapy, it is clear that HIV is the etiologic agent of AIDS.

How Did HIV First Infect Humans?

Two distinct types of HIV can be identified genetically and antigenically. HIV-1 is the cause of the current worldwide pandemic and was first observed in 1984, while HIV-2 was discovered in 1986 and is found primarily in West Africa but rarely elsewhere. Both HIV-1 and HIV-2 are thought to have arisen from simian immunodeficiency virus (SIV). HIV-2 is more closely related to the SIV found in West Africa. Both HIV-1 and HIV-2 have the same modes of transmission and are associated with similar opportunistic infections and AIDS.

In persons infected with HIV-2, immunodeficiency seems to develop more slowly and tend to be milder. HIV-2 infected individuals are also less infectious early in the course of infection. HIV-1 and HIV-2 also differ in geographic patterns of the disease; the United States has few reported HIV-2 cases, but otherwise clinically the diseases are very similar. There are three subgroups of HIV-1—M (main or major), N (new), and O (outlier). Type O HIV-1 is mostly found in Cameroon and Gabon, while the rare N subgroup is also found in Cameroon. In 2009 a new strain, closely related to gorilla simian immunodeficiency virus, was discovered in a Cameroonian woman. It was designated HIV-1 group P. In 2011 a published report confirmed that the virus was again identified in an HIV-seropositive male hospital patient in Cameroon, confirming that the group P virus is circulating in humans. A larger survey revealed that its

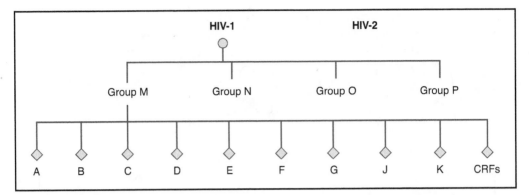

Figure 3-2 Diagram illustrating HIV classification levels. HIV-1 is divided into its representative groups and group M is divided into subtypes and circulating recombinant forms (CRFs).

prevalence was low in the population, accounting for only 0.06% of all HIV infections in Cameroon. The data, however, confirms the mechanism of HIV introduction into the population and suggests that despite its rarity, this newest strain is adapting in the human population.

Scientists believe that SIV might have infected humans on separate occasions to give rise to the three subgroups. Currently, there are approximately 10 different HIV-1 subtypes within group M (Figure 3-1). Subtype B is the major subtype found to infect populations in North America, Latin America and the Caribbean, Europe, Japan, and Australia. Almost all of the M subtypes are found in sub-Saharan Africa. Subtypes A and D are found at the highest rates in central and eastern Africa and C in southern Africa (Figure 3-2). Type C is the predominant form found in India and also causes the most infections worldwide.

Occasionally, two viruses of different subtypes can meet in the cell of an infected person and combine their genetic material creating a new hybrid virus (a process similar to sexual reproduction, and sometimes called "viral sex"). Fortunately, many of these new strains do not survive for long, but those that infect more than one person are known as "circulating recombinant forms" or CRFs. For example, the CRF A/B is a mixture of subtypes A and B in the same individual.

SIV and the Origin of AIDS

Scientists today generally agree that simian viruses resembling the human immunodeficiency virus (HIV) originally spread from chimpanzees and sooty mangabeys to people in central Africa. Most AIDS researchers currently believe that HIV is the virus that causes AIDS and that two different strains of a virus affecting monkeys probably combined to create the form of HIV that has spread around the world. A theory put

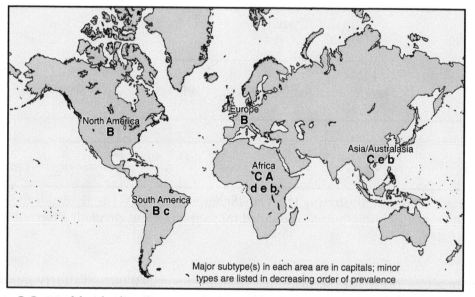

Figure 3-3 Worldwide distribution of HIV subtypes. This map shows the global distribution and genetic diversity of the major subtypes, known as *clades*, of HIV-1.

forward by a team from Nottingham University, who performed an extensive genetic analysis of SIV strains, proposes that wild chimps simultaneously infected with two different SIV strains, which had "viral sex" forming a third virus that could be passed on to other chimps and was capable of infecting humans and eventually causing AIDS (Figure 3-4).

In February 1999, a University of Alabama group led by Dr. Beatrice H. Hahn found a type of SIVcpz that was almost identical to HIV-1 from a frozen sample taken from a subgroup of chimpanzees (Pan troglodytes) once common in west-central Africa. The researchers then collected blood samples from several wild-born baboons and monkeys in Cameroon-17 species altogether. Dr. Hahn and colleagues reported that only chimpanzees in west-central Africa harbor a viral strain that is "truly closely related" to the most lethal AIDS strain. They claimed that this sample proved that chimpanzees were the source of HIV-1 and that the virus at some point crossed species from chimps to humans.

Some have theorized that SIVcpz was transferred to humans as a result of administration of the oral polio vaccine (OPV). It has been known for a long time that the attenuated virus used to make the OPV was grown in monkey kidney cells, further fueling this theory. However, in February 2000, the Wistar Institute in Philadelphia, Pennsylvania, who was involved in the distribution of the polio vaccine found some vials that were frozen since the 1950s. Tests performed by three independent

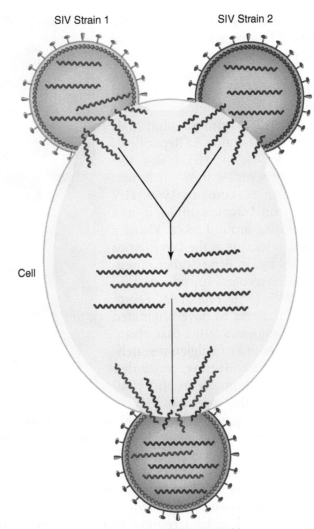

SIV Strain 1

SIV Strain 2

Cell

Third strain of SIV that is highly
pathogenic and can infect humans

Figure 3-4 Schematic representation of the proposed mechanism for the creation of HIV through viral sex between two different SIV strains infecting the same chimpanzee cell. It is likely that the larger chimpanzee fed on smaller monkeys carrying these different SIV strains that later produced a third virus that we now know as HIV.

laboratories on the 1950s-era polio vaccine samples failed to find any traces of SIV, HIV-1, or DNA indicating that chimpanzee cells were used to prepare the vaccine. It was later confirmed that Asian macaque monkey kidney cells were used for the OPV development and not chimpanzee (www.aidsorigins.com/ content/view/121/29/).

Many now believe that SIVcpz was transferred to humans as a result of chimps being killed and eaten or their blood getting into cuts or wounds of the hunters (the Hunter Theory). The earliest known instances of HIV transfer to the human population have also been a point of interest for many. A plasma sample taken in 1959 from an adult male living in what is now the Democratic Republic of Congo tested positive for HIV-1. HIV-1 was also found in archived tissue samples from an American teenager who died in St. Louis in 1969. HIV was also found in the stored tissue samples from a Norwegian sailor who died around 1976. Using a new statistical method, scientists at the Los Alamos National Laboratory, led by Dr. Bette Korber, speculate that HIV-1 might have been introduced into humans around the 1930s (*New York Times*, February 2, 2000). HIV-2 is thought to have originated from the SIV in Sooty Mangabeys rather than chimpanzees. In May 2003, a group of Belgian researchers led by Dr. Anne-Mieke Vandamme, concluded that HIV-2 first appeared in humans around 1940 to 1945. It is possible that the spread of HIV to

Figure 3-5 HIV is believed to have crossed species from monkeys to humans.

humans was made possible because of the global vaccination strategy on the African continent that coincided with colonialism. Reports indicate that African healthcare professionals reused needles on a regular basis as individuals were being vaccinated against a host of infectious diseases that threatened the very existence of the human race at that time. Because needles and syringes were required but expensive, a single healthcare worker might have used a single syringe to inoculate multiple individuals in a given day. This practice could have resulted in the transfer of HIV from infected hunters to the general population. The harsh labor camps, lack of food, poor sanitation, and weakened immune system resulting from colonial rule could have lead to the adaptation HIV if SIV-infected chimps were eaten. The eating of chimps was and still is practiced in many areas of central Africa.

The proponents of the many conspiracy theories implying that HIV was man-made as a bioweapon, mainly to wipe out minorities and homosexuals, ignore several important facts. They first ignore the current evidence that the virus has been identified in people as far back as 1959. At the time when the virus first appeared in the population, we lacked the genetic-engineering technology necessary to "create" the virus. James D. Watson and Francis Crick proposed the idea that the DNA's structure was a double-helix in 1953. It was not until 1958 that the double-helical form of DNA was proven to be correct in the Messelson-Stahl experiment. Moreover, the basic techniques needed to begin the process of creating a virus like HIV, PCR technology, was not discovered until 1983. These theorists also ignore the clear link between SIV and HIV. All evidence therefore suggests that HIV is a recent disease of humanity.

The Africa, Haiti, United States Route of HIV-1 Subtype B

Research completed by Michael Worobey of the University of Arizona and his colleagues suggests that HIV entered the United States via Haiti, possibly arriving in just one person in about 1969, which is earlier than previously believed. Worobey and his colleagues conducted genetic studies using archived blood samples from AIDS patients to construct genetic family trees for HIV. The team obtained and analyzed blood from five of the first AIDS patients identified in the United States. It turns out that all of these patients were recent immigrants from Haiti. The team also analyzed genetic sequences from another 117 AIDS patients that they were able to obtain from around the world who were infected with HIV-1, group M subtype B, the virus strain that has spread most widely. They discovered that the ancestry of most viruses in the United States can be traced back to one common ancestor that came from Haiti. The data suggests that the virus moved from central Africa to Haiti in the 1960s. This goes a long way to explain the early observations that AIDS had a high prevalence in individuals from Haiti. The speculation is that HIV probably arrived in Haiti from the Congo via Haitians who were working in the Congo and returning home in large numbers during those years. It is possible that HIV entered the United States multiple times by single individuals from Haiti before setting off what we now know as the AIDS epidemic. Once in the United States, the virus made its way all over the world through the ease of air travel, making sub type B the most common HIV-1 strain across the world. To see a genetic tree showing that HIV made its way from Africa to Haiti, and then to the United States, visit http://news.nationalgeographic.com/news/bigphotos/87865606.html.

Name: _____

1. It is now generally accepted that HIV-1 originated from _____.
   (A) contaminated polio vaccines
   (B) simian immunodeficiency virus (SIV) in chimpanzees
   (C) bioweapon test gone bad at Fort Dietrich
   (D) homosexual activities that created a new viral strain

2. True or False: HIV is believed to have traveled from central Africa to the United States and then across the rest of the world.

3. True or False: Thabo Mbeki, former South African president, has been accused of causing the death of over 330,000 of his people because of his belief in the HIV/AIDS dissident theory.

4. Name the scientist who is the leading HIV/AIDS dissident (the idea that HIV does not cause AIDS).

5. What is the main theory about how HIV initially came into existence?

6. What is the strongest evidence that HIV is the causative agent of AIDS?

7. Which subtype of HIV is most prevalent in North America, Canada, and Europe?
   (A) A
   (B) B
   (C) C
   (D) E

CHAPTER 4

HIV and the Immune System

Since HIV infection leads to AIDS by destroying the body's defenses, one needs to understand how the immune system functions under normal circumstances in order to appreciate how HIV really works. The immune system is composed of many inter-dependent cell types that collectively protect the multicellular organism from bacterial, parasitic, fungal, or viral infections and from the growth of tumor cells. Typically, the body's immune system vigorously fights infection and disease in a rapid and efficient manner. This is generally accomplished through the numerous cells that make up the major functional components of the system. However, when one part of the system doesn't work correctly, your body becomes vulnerable to many kinds of diseases.

Innate and Adaptive Immune Responses

Humans possess a plethora of immune defensive mechanisms that are employed every day to combat infections. The immune system employs two main types of defense mechanisms when it encounters an antigen (any "foreign" substance) that induces the production of immune defense proteins. The first is the innate or nonspecific defense, which protects the body from various forms of infectious agents. Innate defenses include physical barriers such as the skin, but also employ the acidity of the gastric acids in our stomach, or lysozyme in our tears as well as phagocytic macrophages, neutrophils, and natural killer (NK) cells to combat pathogens. These physical barri-ers are very effective in abrogating the entry of bacterial and viral organisms through destruction of the infectious threat or by physically preventing its attachment to host cells through competition from normal flora. The innate defense system therefore provides rapid response to an invader, but it does not confer long-lasting immunity to the host. The innate system, however, is involved in the recruitment of immune cells to sites of infection, through the production of specialized chemical messengers, called **cytokines** and chemokines. Furthermore, the innate system is directly involved in the activation of the **complement cascade** to identify bacteria, activate cells and to promote clearance of dead cells or **antibody complexes**. Finally, the innate arm of the immune response plays an important role in the activation of the **adaptive immune system** through a process known as **antigen presentation**, where pathogens identified by the innate immune system are engulfed and broken down by specialized cells. Small signature parts of these invaders are presented to cells of the adaptive immune response. This action starts a cascade of events that under normal circumstances results in the clonal expansion of specific lymphocytes and clearance of the invader from the body. Clearly then, the adaptive and innate arms of the immune system work collaboratively to protect our bodies from invaders. The adaptive arm of the immune system is characterized by the specific or directed action of specialized lymphocytes and specific antibody production directed against a particular pathogen or pathogen protein (epitope). The basis of the specificity of the adaptive system lies in the capacity of immune cells called lymphocytes to distinguish between proteins produced by the body's own cells ("self" antigens), and proteins produced by invaders or cells under the control of a virus ("non-self") antigen. The cells of the immune system can engulf bacteria, kill parasites or tumor cells, or kill virus-infected cells. All of the cells that function in the immune system are called white blood cells (WBC).

The WBCs of the immune system are produced in the bone marrow and are then carried in the peripheral blood to specialized organs of the immune system, where they are educated to differentiate between self and non-self in preparation to launch immune responses against infections. These cells originate from the lymphatic system organs, which include the thymus, lymph nodes, and bone marrow. As far as scientists have determined, each cell plays a specific role in fighting disease. Every germ that invades the body has unique identification marks on its surface. Some white blood cells, like macrophages, engulf and destroy bacteria and damaged cells. In fact, many organs of the body have specialized macrophages that play important roles in maintaining a pathogen-free environment. These include kupffer cells in the liver, glial cells in the brain, alveolar macrophages in the lungs, osteoclasts in the bones, Hofbauer cells in the placenta, and histiocytes in connective tissues.

B-lymphocytes, commonly referred to as B cells, mature in the bone marrow. Upon activation, they differentiate into memory B cells and plasma cells. They later produce antibodies (host defense proteins; see Figure 4-2 on page 39) with the help of T-lymphocytes (T cells) called T-helper cells. T-lymphocytes mature in a lymphoid organ called the thymus. The antibodies that are produced can neutralize viruses, bacteria, or toxic proteins in the blood and other body fluids, thereby preventing disease. They can also enhance the phagocytic clearance of microbial antigens by a process called *opsonization*, whereby the specific microbe is coated with antibodies leading to more efficient uptake and disposal.

Central Role of CD4 Cells

Picture the cells of the immune system as a group of well-trained Navy Seals and rangers in the heat of battle. Because of the terrain in which they find themselves, the members of this outfit cannot directly see or talk to each other, but are organized and directed by a commanding officer who is constantly sending each solider a stream of messages with the GPS coordinates of the enemy as well as his fellow combatants. In the human body, the role of the commanding officer would be filled by the T-helper cell (also known as the CD4 cell because of the major receptor on its surface). The T-helper cells are therefore central to the activity of both the cell-mediated and antibody-producing arms of the adaptive immune response. All of the cells needed to mount an effective adaptive immune response require help from the T-helper cells. Under normal circumstances, CD4 is an important co-receptor for T cell activation in the process of identifying and eliminating an invader. Specifically, CD4 is a co-receptor that assists the T cell receptor (TCR) in communicating with an antigen-presenting cell. It was never designed for HIV to gain entry into these cells. The virus has just evolved to exploit this very important protein molecule found on the surface of some very important immune cells.

The CD4 cell dictates the behavior of the other immune cells almost from the very moment of an infection. The major types of cells controlled by the CD4 cell include the APCs, which are at the front line of the immune system. These cells are constantly patrolling the tissues and are responsible for identifying foreign entities by recognizing proteins, or antigens, on the surface of an invader, which marks it for death. The APCs then engulf the foreign antigen, break it down, and display portions of it on their surfaces (Figure 4-1). These pieces that are displayed serve as a signal to the CD4 cell that the body is under attack, and it needs to mobilize the other members of the outfit. **B-cells** differentiate into plasma cells, which produce **antibody** molecules (Figure 4-2) that are specifically designed to tackle and "lock" onto antigens on the surface of virus or bacterial particles to prevent them from replicating.

These B cells are activated by chemical signaling proteins called *cytokines*, produced by the CD4 cells, causing the B cells to produce antibodies to the specific threat (Figure 4-4). The antibody molecules, which are Y-shaped, attach themselves to the virus or bacteria and prevent them from causing infection. Antibodies bound to the surface of invaders also serve as a signal for macrophages, which act like scavenger cells that find and engulf any bacteria or virus particles marked by the antibodies. CD4 cells also lead to the activation of killer T cells (Figure 4-3) that hunt for viral antigens, which indicate that host cells have been infected with a virus and turned

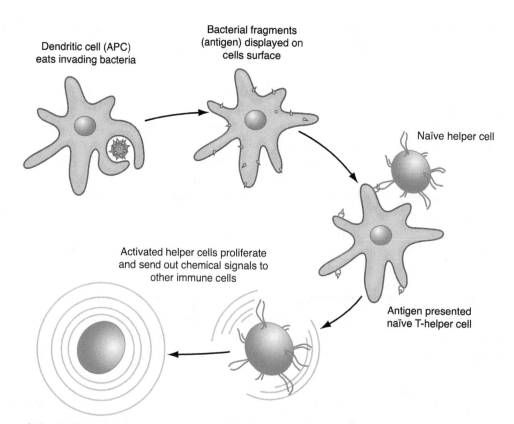

Dendritic cell (APC)
eats invading bacteria

Bacterial fragments
(antigen) displayed on
cells surface

Naïve helper cell

Activated helper cells proliferate
and send out chemical signals to
other immune cells

Antigen presented
naïve T-helper cell

Figure 4-1 Antigen presentation by a dendritic cell (APC) to T-helper cells. The APC engulfs the bacteria and breaks it down into fragments that are presented on the surface of the cell to naïve T-helper cells. These naïve cells become activated, secreting chemical messengers that "call" in other immune cells to the site as well as help to activate B cells and cytotoxic T cells.

into a virus-producing machine. The killer T cell, as its name suggests, kills the infected cell to stop it from producing more virus particles. Within a few days, the immune system can eliminate a typical flu or cold virus infection. Once a virus is eliminated, however, the body doesn't forget about it. Some of the B cells and T cells that were mobilized for the fight become memory cells that stay in the body for many years. If the same virus comes back, those cells will quickly find it, clonally expand to increase the number of responders and eliminate it, before you ever become sick. That is why some diseases, like the measles, only strike once in a lifetime. It is not that you never encounter the causative agent again; however, when you do the second, third, etc. time around, your immune system just has to consult its vast database for sensitized cells that are ready to combat the infectious agents before it is able to cause disease pathology. This is the principle upon which vaccination is built.

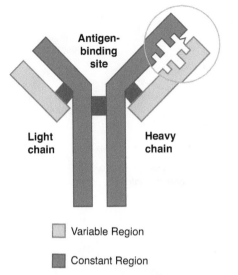

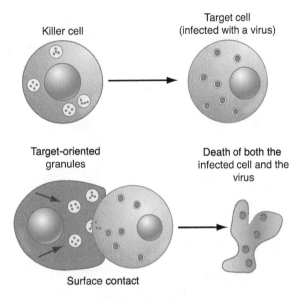

Figure 4-2 **Antibody molecule**. An antibody is made up of two heavy chains and two light chains. The variable region, which differs from one antibody to the next, allows an antibody to recognize its matching antigen.

Figure 4-3 Killer T cells directly attack other cells carrying certain foreign or abnormal molecules on their surfaces.

How Does HIV Attack the Immune System?

HIV, like the other members of the lentivirus family of retroviruses, is unique among viruses because it directly attacks the CD4 helper T cell, the very heart of the immune system. By targeting this irreplaceable cell, the virus interrupts the entire immune response, leaving the host vulnerable to any infectious agent. Once inside the body, HIV hunts down, attaches to, and penetrates the CD4 cell. Remember that the other cells need the CD4 cell as their commander to tell them what they need to do to rid the body of the invader (see Figure 4-4).

HIV, however, hijacks the CD4 cell and uses its machinery to make virus particles. Although some immune cells, including NK cells and cytotoxic T-lymphocytes, recognize and destroy HIV particles, the virus readily attaches to the CD4 receptor of the helper T cells and enters the cell where it starts a productive infection. Since cells of the immune system are constantly circulating in and otut of the lymphoid organs, the HIV-infected cell will soon carry the virus into areas of the lymphatic system, where CD4 cells are most abundant, including the 500 to 600 lymph nodes all over the body

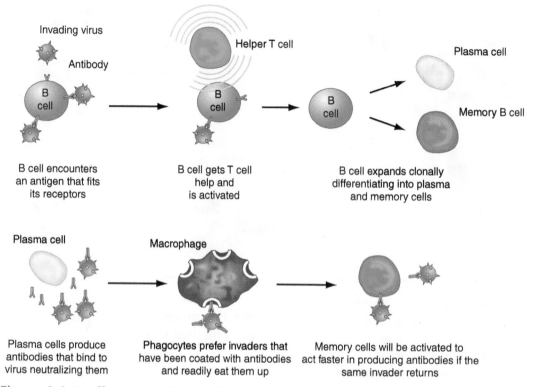

Figure 4-4 B cell activation by specific antigen and T cell help. This figure outlines the very important role played by the T-helper cells in the immune response. Without T cell help, the B cell does not expand and differentiate.

(see Figure 4-5 for lymph node regions). Once there, HIV begins replicating and infecting CD4 cells more rapidly, replicating itself billions of times each day.

Cellular Receptors and Entry of HIV

As illustrated in Figure 2.2, HIV resembles a sphere with spiky proteins all over its surface. In order to infect a CD4 cell, one of those proteins, known as gp120, binds to a receptor protein, CD4, on the T cell's surface. A second protein, called gp41, then locks onto a second receptor protein, CCR5 or CXCR4 as previously discussed. CCR5 is predominantly expressed on T cells (memory and activated CD4 lymphocytes), gut associated lymphoid tissues (GALT), macrophages, dendritic cells, and microglia. CXCR4 is expressed on T cells (naïve and resting CD4 lymphocytes, as well as CD8 cells), B cells, neutrophils, and eosinophils. Up to 90% of newly transmitted HIV uses the CCR5 coreceptors to enter the cells and are said to be M-tropic. Recent research has

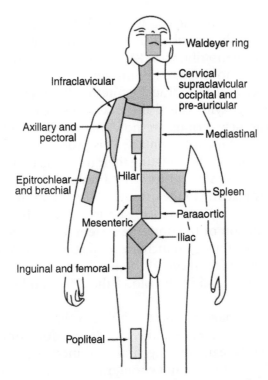

Figure 4-5 The average human has approximately 500–600 lymph nodes distributed throughout the body. Major clusters can be found in the underarms, groin, neck, chest, and abdomen.

demonstrated that CCR5 viruses are more often transmitted sexually. CXCR4-tropic viruses emerge in approximately 50 to 60% of infected individuals within five years after initial contact. CXCR4-tropic (T-tropic) viruses are associated with pronounced depletion of CD4 T cells, correlate with syncytia formation, and are associated with more rapid progression to AIDS. This is because CXCR4 is expressed on nearly all CD4 T cells, whereas only 15 to 30% express CCR5. Therefore, CXCR4 viruses have a wider range of susceptible target cells, leading to higher viral load, lower CD4 T cell counts, and an overall reduction in immune response capabilities. These coreceptors are therefore very important in HIV infection. A tremendous amount of resources are now being placed into performing HIV phenotyping tests to determine the tropism of the virus in an individual patient. This has implications for treatment as well since FDA approved anti-HIV drugs like Selzentry and Vicrivorac are CCR5 or entry inhibitors. These drugs are effective against CCR5 tropic viruses but have no effect on CXCR4 tropic ones.

Once both connections are made, the virus can fuse, penetrate the outer membrane of the cell, and insert its viral proteins and RNA.

Once inside the cell, the virus hijacks the cellular machinery, transforming it into an HIV factory. As previously described under HIV replication, several enzymes are employed by the virus to transcribe viral RNA into DNA, to integrate the viral DNA into the cellular DNA, and to activate the provirus. After the copies are made, they gather at the cell's surface and form buds before finally disconnecting from the host cell. Once they're free, an enzyme called the viral protease cuts the new proteins into small, usable pieces. The new virions are now mature and ready to infect new cells with which they will soon come into contact. This process of infection and viral budding destroys the CD4 cells, slowly depleting their numbers (Figure 4-6). Many people experience flu-like symptoms, sometimes with a rash, two to three weeks after being infected. This is the body reacting to the invading HIV and mounting a strong immune response. At this point the body is producing up to a billion helper and killer T cells a day. The constant stimulation of the immune system leads to activation of CD4 cells that are already infected, leading to an increase in the production of new viruses. Those viruses in turn go out and infect more of the newly created CD4 cells. At this point, there is a large enough viral load in the blood that HIV can be diagnosed through a blood test.

Soon, the extensive damage to these T-helper cells takes its toll and the total number of these cells begins to fall steadily. As destruction continues, viral particles are released into the bloodstream and the viral load increases steadily. Over time, the antibody levels will fall since there is not enough T cell help for the HIV-specific B cells to continue producing these defense proteins. The reduction in CD4 cell count with a simultaneous reduction in protective antibodies signals the progression of HIV infection and signs of immunodeficiency soon become visible.

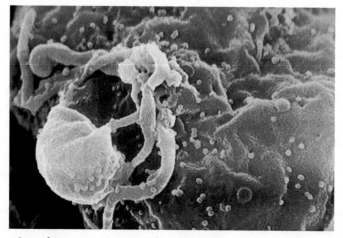

Figure 4-6 Scanning electron micrograph of HIV-1 budding from cultured lymphocyte. Multiple round bumps on the cell surface represent sites of assembly and budding of virus particles.
(Source: Content providers: CDC/ C. Goldsmith, P. Feorino, E. L. Palmer, W. R. McManus. http://phil.cdc.gov/ PHIL_Images/10000/10000.tif)

Name: _____

1. HIV specifically attacks and kills helper T cells. This detrimentally weakens the immune system because helper T cells _____.

   (A) routinely produce antibodies

   (B) can effectively turn off the immune response

   (C) directly destroy cells infected with HIV

   (D) activates the cell-mediated and antibody-mediated immune defenses

   (E) are responsible for roaming the body in search of dangerous invaders

2. Over time, HIV infection leads to the depletion of immune cells through a variety of mechanisms; one of which is mediated by the formation of syncytia. _____ viruses correlate with syncytia formation that takes place _____.

   (A) M-tropic, only in uninfected circulating T-lymphocytes

   (B) CXCR4 tropic, when Gp120 on infected cell binds CD4 on the surface of uninfected cells

   (C) CXCR4-tropic, only in individuals infected with HIV-2

   (D) CXCR5-tropic, when HIV particles cling together forming a large mass of over 50 viral particles

3. In the initial events of HIV infection the _____ molecule on the surface of the virus binds to the _____ receptor present on susceptible cells.

   (A) CD4, p24

   (B) Gp120, CD4

   (C) CD4, Gp41

   (D) Gp41, CD4

   (E) Gp160, CCR5

4. Which of the following immune response cells directly makes and secretes antibodies?

   (A) Helper T-cells

   (B) B-cells

   (C) Plasma cells

   (D) Natural killer cells

   (E) CD8 T-cells

5. True or False: The innate and adaptive immune responses work independently of each other.

6. True or False: Macrophage cells that live in the liver are called kupffer cells.

7. Why are CXCR4-tropic (T-tropic viruses) associated with more pronounced depletion of CD4 T cells?

8. What is the role of the CD4 co-receptor in a regular immune response?

CHAPTER 5

Progression of HIV
Infection to AIDS

The battle between this resilient retrovirus and the immune system cells can continue to rage in the body for many years and even decades, since the two entities have such a large supply of ammunition. During this time, the HIV-infected person might feel completely healthy. The infected individual can, however, spread HIV to anyone with whom he or she exchanges bodily fluids, such as blood, vaginal fluid, semen, or breast milk, because the virus is active. When an infected cell starts producing HIV proteins, the HIV envelope proteins migrate to the cell membrane and are displayed on the cell surface, just as if they were poking out of an HIV particle. To other uninfected, CD4-bearing cells, the infected cell now looks like an HIV particle and is therefore bound by these cells. Therefore, an infected T-helper cell can join with a healthy, uninfected T-helper cell in a similar manner as the HIV particle that originally attached to and entered into that host cell. This is repeated until eventually you have one large HIV-infected CD4+ cell with as many as 50 nuclei called a syncytium (Figure 5-1). Syncytia do not always form in HIV-infected patients (about 50% of people after five years), however, when it does, it serves as a route for rapid CD4 cell depletion since these bound cells cannot function in immune protection. Individuals with syncytia progress to AIDS more rapidly and are more likely to have neurological involvement.

A major reason that HIV is unique is the fact that despite the body's aggressive immune responses, which are sufficient to clear most viral infections, some HIV invariably escapes. This is due in large part to the high rate of mutations that occur during the process of HIV replication. Even when the virus does not escape the immune system by mutating, the body's top soldiers in the fight against HIV—subsets of killer

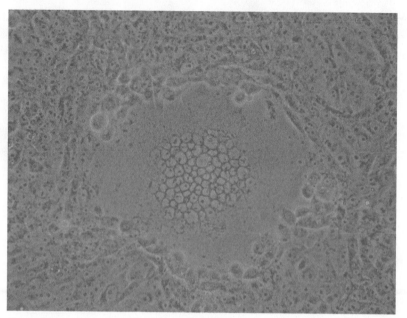

Figure 5-1 Syncytium "forming"-type cytopathic effect induced by the virus infection. Phase-contrast microscopic image of 100-fold magnification.

T cells that recognize HIV—may be depleted or otherwise become dysfunctional. Eventually, HIV gains the upper hand and begins to kill off more CD4 cells than the body can produce. Without the CD4 cells, the B cells and killer T cells lack instruction about where to go and what to look for, and the body's immune response becomes less and less effective over time. The viral load in the plasma quickly rises and the amount of antibodies produced by the B cells fall. This leaves the patient with no real protection against the weakest of pathogens.

Stages of HIV Infection

In addition to the serious opportunistic infections that characterize the AIDS stage of HIV infection, there may be several other signs and symptoms of the disease at various stages of the infection process. It is not unusual for people with AIDS to have systemic symptoms of infection, including fevers, chills, night sweats, swollen lymph glands, generalized weakness, and severe weight loss. Using the CD4 cell counts, HIV antibody levels, and the opportunistic infections associated with HIV infection, the entire infection process can generally be broken down into four distinct stages: **primary infection, clinically asymptomatic stage, symptomatic HIV infection,** and **progression from HIV to AIDS.**

The **primary or acute stage** of infection lasts for only a few weeks and is often accompanied by short flu-like symptoms and a variety of other signs and symptoms, which vary based mainly on the pre-infection health of the patient (Figure 5-2). Almost 20% of those infected feel sick enough to consult a doctor at this stage of the illness. At the beginning of the primary stage the viral load increases rapidly, inducing an immune response. At this point, an antibody test would be negative because the immune system needs time to respond and make HIV-specific antibodies. As antibody production increases (seroconversion) and killer T cells start responding to the virus, the viral load begins to fall because the antibodies and T cells keep them in check. The period between infection and seroconversion is referred to as the **window period** because all antibody tests will be negative at this stage. However, if nucleic acid tests are done at this time they will come up positive for the presence of viral RNA.

The **clinically asymptomatic (latent) stage** of HIV infection is very characteristic of lentiviruses of which HIV is a family member. This period last an average of 10 years and may be quite uneventful, being free from major symptoms, although there may be swollen glands.

The level of HIV particles in the peripheral blood falls to very low levels, but people remain infectious throughout this stage and HIV antibodies are detectable in the

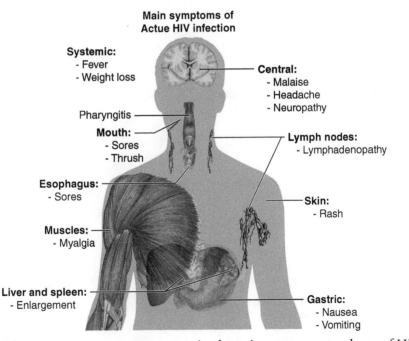

Figure 5-2 Most common symptoms seen in the primary or acute phase of HIV infection.

blood. It should be noted that HIV is by no means dormant during this stage, but is very active in the lymph nodes and other areas where CD4 cells are plentiful. Antibody tests will be positive at this stage and CD4 cell counts will start to fall.

As time progresses, the lymph nodes and other tissues become damaged because of the years of active HIV infection. The virus mutates and becomes more virulent, leading to increased T helper cell destruction. Despite its best efforts, the body fails to keep up with replacing the T helper cells that are lost, and the CD4 levels fall precipitously as reflected in peripheral blood counts. As the immune system continues to fail, symptoms begin to manifest. Initially, many of the symptoms are mild, but as the immune system deteriorates the symptoms worsen. This is the symptomatic stage of HIV infection, which is mainly caused by the emergence of opportunistic infections and cancers that the immune system would normally prevent. These can occur in almost any organ system of the body (Figure 5-3). The most common opportunistic infections are discussed below. It should however, be noted that the latent stage of the disease can be prolonged indefinitely with the right treatment combination, which is designed to keep the viral load down and the CD4+ T cells at normal levels.

Progression of HIV to AIDS is the last stage of the disease. At this stage the body becomes overwhelmed with opportunistic infections as more CD4 cells are lost. The CD4 cell count drops below 200 and the patient feels very ill. Most of the day is spent

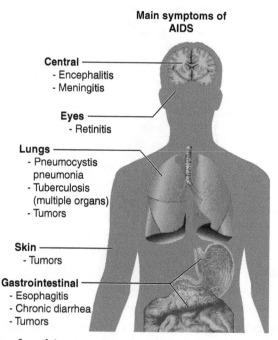

Main symptoms of AIDS

Central
- Encephalitis
- Meningitis

Eyes
- Retinitis

Lungs
- Pneumocystis pneumonia
- Tuberculosis (multiple organs)
- Tumors

Skin
- Tumors

Gastrointestinal
- Esophagitis
- Chronic diarrhea
- Tumors

Figure 5-3 The onset of multisystem involvement and certain opportunistic diseases signify the onset of AIDS.

sleeping or resting, and there might be profound weight loss, leaving the patient look-ing emaciated. Studies suggest that HIV also destroys precursor cells that mature to have special immune functions, as well as the microenvironment of the bone marrow and the thymus that is needed for developing such cells. These organs probably lose the ability to regenerate, further compounding the suppression of the immune system.

When Is a Person Diagnosed with AIDS?

Untreated HIV disease is characterized by a gradual deterioration of immune function. Two to four weeks after exposure to the virus, up to 70% of HIV-infected people suffer flu-like symptoms related to the acute infection. With the persistent depletion of CD4 cells, the immune system loses its ability to protect the infected patient from infections that would ordinarily be easily cleared. A healthy, uninfected person usually has 1,000 to 1,500 CD4+ T cells per cubic millimeter (mm^3) of blood. If an individual is infected with HIV and is not taking antiretroviral medication, the number of these cells in a person's blood progressively declines (Figure 5-4).

When the CD4+ T cell count falls below 200/mm^3, the infected person will become very susceptible to several opportunistic infections and cancers that are characteristic of immunodeficiency or AIDS, the end stage of HIV infection.

AIDS patients often suffer from and later succumb to infections of the lungs, intes-tinal tract, brain, eyes, and other organs (Figure 5-3). This is typically accompanied by debilitating weight loss, diarrhea, neurological disorders, and certain rare cancers such as **Kaposi's sarcoma** and certain types of lymphomas. There is widespread HIV-mediated destruction of the lymph nodes as well as the spleen and other organs of the

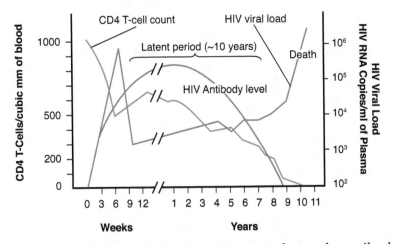

Figure 5-4 Progression of HIV infection to AIDS. Note that as the antibody titer increases, the viral load is decreased. As the CD4 count falls, so does the antibody level and the viral load again soars.

immune system. This leads to further immunosuppression seen in people with AIDS. Immunosuppression by HIV is confirmed by the fact that medicines, which interfere with the HIV life cycle, result in the preservation of CD4+ T cells and immune function as well as delay clinical illness. HIV infection leading to AIDS, however, is not uniformly manifested in all individuals. A relatively small proportion of individuals infected with HIV rapidly develop AIDS and die within months after primary infection. At the same time, approximately 5% of HIV-infected individuals exhibit no signs of disease progression even after 15 or more years. Antiretroviral drugs can significantly prolong the time between initial HIV infection and the onset of AIDS. Modern combination therapy is very effective and, theoretically, someone with HIV can live for a very long time before progressing to AIDS. Unfortunately, millions of AIDS patients in many developing parts of the world cannot access or afford these medications and therefore continue to die. The good news is that there continue to be steady progress in this area resulting mainly from increased political and economic commitment, stimulated by people living with HIV as well as various drug companies who have partnered with governments and foundations all over the world. As a result, more than 4 million people in low- and middle-income countries were receiving HIV antiretroviral therapy at the close of 2008. It is believed that various host factors such as the virulence level of the individual, HIV strain, age of the person at the time of infection, genetic differences among individuals, immune status, as well as co-infection with other microbes may determine the rate and severity of HIV disease expression from person to person.

AIDS Opportunistic Infections

Each of us carries in our bodies an array of germs—viruses, bacteria, fungi, and protozoa—that exist in a balanced state with our immune system. When our immune system is working well, it controls these microbes. However, when the immune system becomes compromised by HIV, other bacterial and viral diseases, or even by some medications, these germs can grow out of control and cause health problems. HIV does not kill its victims directly; instead, it weakens the body's ability to fight diseases by destroying our T helper cells as explained earlier. Consequently, infections that rarely occur in individuals with a properly functioning immune system can prove deadly to HIV-infected individuals. These infections are collectively termed **opportunistic infections** because they take advantage of the opportunity offered by a weakened immune system to grow in the host at a higher level than they are accustomed to being present, and must be treated if the patient hopes to survive. The CDC has developed a list of opportunistic infections signifying that an individual who is HIV positive has progressed to AIDS (www.hivandhepatitis.com/hiv_ois_list.html). When an HIV-positive person has both a low CD4 count and what doctors call an "AIDS-defining illness" or "opportunistic infection," that person has AIDS. Early symptoms of worsening HIV infection include night sweats, weight loss, fatigue, and swollen lymph nodes.

Common Opportunistic AIDS Infections

Some of the most common opportunistic infections affecting HIV/AIDS patients are:

1. **Kaposi's sarcoma (KS)**—A type of skin cancer that presents with purple lesions all over the body (Figure 5-5). Before the advent of AIDS, KS was seen only rarely in older men, mostly of eastern European descent. Although KS was one of the defining signs of the disease in the early years of AIDS, with current treatment regimens, KS is less commonly seen than it was during the 1980s and early 1990s.

2. **Pneumocystis carinii pneumonia (PCP)**—A rare form of pneumonia typically seen in people who are severely immunosuppressed. The main symptoms are shortness of breath and difficulty breathing, as the lungs fill up with fluid (Figure 5-6). Treatment requires heavy doses of antibiotics combined with oxygen therapy. PCP continues to be a common infection in people with AIDS, and many suffer several bouts of this serious infection before dying.

3. **Toxoplasmosis**—An infection caused by a parasite, *Toxoplasma gondii*, that infects the brain and can cause brain damage and blindness. The infection affects 10% to 20% of people in North America by the time they are adults, but the immune system normally keeps it under control. Before AIDS, toxoplasmosis was considered a minor danger only for pregnant women when changing a cat litter box or working in an area contaminated with cat feces. Infection in the unborn child early in pregnancy can result in miscarriage, poor growth, early delivery or stillbirth. If a child is born with toxoplasmosis, he or she can experience eye problems, hydrocephalus (water on the brain), convulsions, or mental disabilities. Toxoplasmosis has become a serious infection in AIDS patients and may infect the eyes, brain, and heart muscle cells (Figure 5-7).

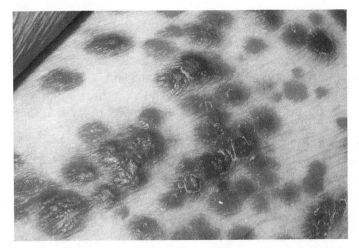

Figure 5-5 The purplish lesions of Kaposi's sarcoma in an AIDS patient.

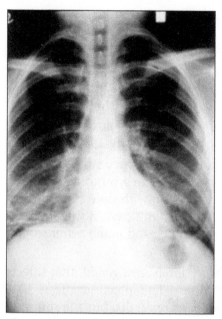

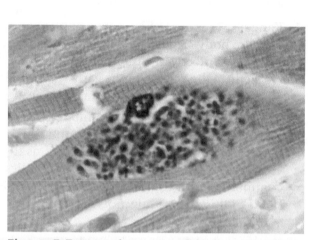

Figure 5-7 Toxoplasmosis of the heart in AIDS. Histopathology of active toxoplasmosis of myocardium (heart muscle). Numerous tachyzoites (asexual stage of rapid growth) of *Toxoplasma gondii* are visible within a pseudocyst in a myocyte (muscle cell). Parasite. CDC/ Dr. Edwin P. Ewing, Jr.
(Source: http://phil.cdc.gov/PHIL_Images/966/966.tif)

Figure 5-6 Lung x-ray of a patient diagnosed with AIDS shows PCP infection. There is increased white (opacity) in the lower lungs on both sides, characteristic of *Pneumocystis* pneumonia.

4. **Candidiasis**—A common opportunistic infection in people with HIV. It is an infection caused by a common type of yeast (fungus) found in most people's bodies. A healthy immune system keeps it under control. However, at the point of immunosuppression, candidiasis runs rampant, infecting the mouth, throat, windpipe, skin, or vagina. Candidiasis can occur months or years before other, more serious opportunistic infections. It appears most frequently in men as a white coating of the mouth (Figure 5-8) and causes burning, a bad taste, and lack of appetite. Women often experience this infection as chronic vaginal candidiasis that does not respond to treatment. This infection can infect organs and tissues throughout the body becoming life threatening.

5. **Cytomegalovirus (CMV)**—Another virus that is common in adults, but is normally held in check by a healthy immune system. Between 50 and 85% of the U.S. population tests positive for CMV by the time they are 40 years old. CMV is

a member of the herpes family of viruses and currently one of the leading causes of blindness and death in people with AIDS. CMV can infect any part of the body, but in people with AIDS, it generally attacks the eyes and lungs (Figure 5-9).

6. **Tuberculosis (TB)**—TB is unique among HIV-related infections because it can infect people who are not infected with HIV or are immunocompetent. It is a respiratory pathogen that was on the decrease in most parts of the world until the 1980s when the AIDS epidemic started. TB is transmitted through a respiratory route but is treatable once identified. This is one of few infections that may occur in early-stage HIV disease. However, multidrug resistance is a potentially serious problem. Although the incidence of HIV-associated TB has declined over the past ten years because of drug therapy and other improved practices in Western countries, it remains a serious problem in developing countries where HIV is most prevalent. In early-stage HIV infection (CD4 count >300 cells per μL), TB typically pres-

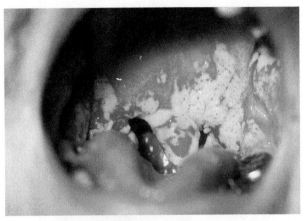

Figure 5-8 Oral thrush or candidiasis. Note the opaque areas on the gum lining outlined by the circle.
(Source: CDC http://phil.cdc.gov/PHIL_Images/02112002/00018/PHIL_1217.tif)

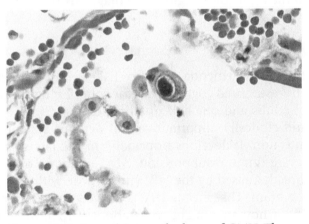

Figure 5-9 Lung histopathology of CMV. The central cell displays the dramatically enlarged intra-nuclear inclusion, characteristic of CMV.
(Source: CDC http://phil.cdc.gov/PHIL_Images/958/958.tif)

ents as a pulmonary disease (Figure 5-10). However, in advanced HIV infection, TB often presents with systemic disease affecting other extra pulmonary organs. Symptoms are usually not localized to one particular site and may involve infection of the bone marrow, urinary and gastrointestinal tracts, liver, regional lymph nodes, and the central nervous system.

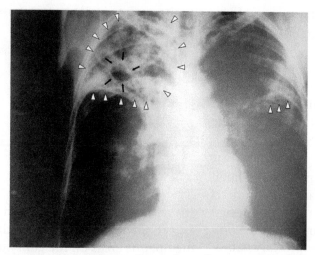

Figure 5-10 An anteroposterior x-ray of a patient diagnosed with advanced bilateral pulmonary tuberculosis. This AP x-ray of the chest reveals the presence of bilateral pulmonary infiltrate (white triangles), and "caving formation" (black arrows) present in the right apical region. The diagnosis is advanced tuberculosis.

AIDS Dementia Complex

The AIDS dementia complex (ADC), also known as HIV-1-associated dementia or HIV-associated cognitive/motor complex, has been estimated to affect up to one-third of adults and one-half of all children with AIDS. ADC is one of the most common and clinically important central nervous system (CNS) complications of late HIV-1 infection. It develops principally in the late stage of HIV-1 infection and associated severe immunosuppression. ADC might be one of the only AIDS-related complexes directly caused by the HIV virus. Those with ADC have HIV-infected macrophages in the brain. That means HIV is actively infecting brain cells. There are several theories about how this might come about. The most common explanation currently is that HIV enters the brain via HIV-infected monocytes, which then differentiate into macrophages. The virus replicates in these cells and can then, in theory, infect other cells in the immediate surroundings; macrophages and microglial cells are the most common. These infected cells are thought to secrete chemical messengers and inflammatory proteins that lead to neurotoxicity. The term *dementia* is used by neurologists to describe a clinical syndrome composed of memory loss, decreased mental concentration, and the loss of other intellectual functions because of progressive disease in the brain. There may be motor signs, such as weakness, lack of coordination, and unsteady gait. Other symptoms may include loss of interest in one's surroundings, as well as severe mobility problems. Prior to effective antiretroviral therapy, ADC occurred in more than 60%

of patients who developed AIDS. With the use of highly active antiretroviral therapy (HAART), the incidence has declined to about 10% to 25% in Western countries, however the prevalence in India is only 1 to 2%. Some have speculated that this difference might be due in part to the HIV sub-type difference seen between the two regions. Almost 99% of HIV-1 infections in India are by subtype C, while many of the other Western regions have subtype B. Published research seems to suggest that subtype C has evolved to be less pathogenic to the human host. ADC is often treated with antipsychotics, antidepressants, psychostimulants, antimanics, and anticonvulsants. These drugs do not treat the underlying cause of ADC, or even stop its progression. However, they may ease some of its symptoms.

Test Your Knowledge

Name: _____

1. As HIV infection progresses out of latency into AIDS, which of the following would be a probable observation?

   (A) CD4 cells will proliferate and increase in number.

   (B) CD8 cells will become undetectable.

   (C) R5 tropic viruses will decrease and X4 tropic viruses increase.

   (D) Viral load will increase.

   (E) Both C and D.

2. AIDS may be accompanied by several opportunistic infections and complexes. Which of the following AIDS-related diseases is caused directly by the HIV virus?

   (A) AIDS dimentia complex

   (B) Lymphoma

   (C) Severe bacterial pneumonia

   (D) Oral candidiasis

3. Several infections are associated with later stage HIV infection. Which of the following infections can infect the respiratory tract of HIV patients and was common in AIDS patients in the 1980s?

   (A) Toxoplasmosis

   (B) Isosporiosis

   (C) Pneumocystis carinii pneumonia

   (D) Virucella zoster

4. An HIV-infected person is diagnosed with AIDS _____.

   (A) when his/her CD4 cell count is equal to 500

   (B) after 10 years of infection

   (C) when his or her immune system is seriously compromised

   (D) when his or her CD8 cell count is below 200

5. Which of the following combinations would you expect to find in an individual who has been diagnosed with AIDS?

   (A) CD4 cell count below 200 and high viral load
   (B) Low levels of HIV antibodies and low viral load
   (C) Low viral load and high levels of Gp 120 antibodies
   (D) High CD4 count and high viral load
   (E) High CD4 count and opportunistic infections

6. What is the window period of HIV infection?

7. What does the term seroconversion mean in the context of HIV infection?

CHAPTER 6

Transmission of HIV

With the absolute devastation that AIDS has brought upon the world's population, you might be surprised to learn that compared to other viruses such as hepatitis A, which can remain active for an extended period of time outside the body, HIV is a very fragile virus. The fact is, outside the body, HIV cannot withstand the many environmental pressures that most organisms have to contend with in order to survive. Factors such as heat, desiccation or drying, and various chemicals prove detrimental to the virus. Therefore, HIV is spread through direct contact with the body fluids of an infected person and cannot be contracted by casual contact. Although HIV has been found in tears, sweat, and saliva, the number of viral particles in these fluids is very low. Health officials maintain that the numbers in these fluids are too low to result in transmission of a productive infection. HIV can be transmitted through blood and body fluids because they contain cells that can be infected with the virus or the cell free viral particles can survive in body fluids and are typically in high numbers.

Major Routes of HIV Transmission

The main routes of HIV infection are (1) penetrative unprotected sexual intercourse with someone who is infected with the virus, (2) transfusion of contaminated blood or injection of contaminated blood products, (3) organ transplants or skin grafts taken from someone who is infected, (4) artificial insemination with infected semen, (5) transmission from an infected mother to her baby during pregnancy, at the time of birth or through breastfeeding, and (6) sharing of an unsterilized needle that was previously used by someone who is infected. Infection through contaminated blood and blood products is currently very rare since blood and blood products are routinely

screened for HIV antibodies or nucleic acid material. There has been only one documented instance of a healthcare worker transmitting HIV to patients in the United
States. This was an unfortunate case where one infected dentist transmitted HIV to six
of his patients during surgery.

Worldwide, the virus is acquired most commonly through infected semen and
blood. Sexually transmitted HIV worldwide is more prevalent in heterosexual relationships. In the United States, however, homosexuals and injection-drug users
account for the highest proportion of HIV-infected persons. HIV and AIDS have had
a tremendous impact on men who have sex with men (MSM) in the United States.
Although MSM represent only 4% of the male population in the United States, in
2010 they accounted for 78.2% of new infections among men and 63% of all new
HIV infections (see Figure 6-1). The estimated number of new HIV infections among
MSM in 2010 increased by 12% from the new infections among MSM in 2008. Currently, MSM account for 52% of all people living with HIV infection, suggesting that
the population that pioneered AIDS awareness in the United States and started a
movement worldwide in the 1980s, might need to again take drastic steps to stem the
tide of HIV/AIDS that appear to be making a strong return especially among young,
gay African American males.

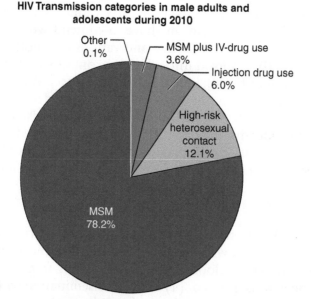

Figure 6-1 HIV transmission by categories for male adults and adolescents in
the United States during 2010. MSM accounted for over two-thirds of all male
transmissions.
(Source: This graph was created based on HIV/AIDS statistics from the CDC. www.cdc.gov/hiv/resources/factsheets/us.htm)

Having another sexually transmitted infection (STI) may also impact one's susceptibility to HIV infection. If an HIV-infected person has another STI, the presence of this STI can increase the risk of HIV transmission. If an HIV-negative individual has an STI, it can also increase their risk of being infected with HIV. This may be because most STIs (e.g., syphilis or herpes) cause the formation of genital warts or ulcers, which could bleed or which can allow the transfer of infected body fluids directly into the bloodstream. Intact, healthy skin is an excellent barrier against HIV, other viruses, and bacteria. Infection with other STIs such as *Chlamydia* or gonorrhea may cause localized inflammation leading to an influx of immune cells in the genital area, thus making HIV transmission much more likely. HIV transmission, however, is more likely in individuals with genital ulcers than in those with an STI that does not cause ulcers. Using condoms during sex is the best way to prevent the sexual transmission of diseases, including HIV.

High-Risk Behaviors

The CDC estimates that approximately 4 to 5 million Americans participate in activities that put them in danger of coming in contact with infected blood or other body fluids. Anyone who participates in these high-risk behaviors is at increased risk of contracting HIV. These activities include sharing needles or syringes; having sexual contact, including oral, with an infected person without using a condom; or having sexual contact with someone whose HIV status is unknown. HIV is frequently spread among injection-drug users by the sharing of needles or syringes contaminated with very small quantities of blood from someone infected with the virus (Figure 6-2). In many parts of the world, often because it is illegal to possess them, injecting equipment are shared. A very small amount of blood can transmit HIV, and can be injected directly into the bloodstream along with the drugs. Crystal methamphetamine use has been shown to double the risk of HIV infection in any single sexual encounter. This may be because users are more likely to be unsafe; however, research is also showing that the body of a person high on meth suffers physiological changes that make transmission easier.

Female prostitutes often have 200 to 300 sexual partners per year and are therefore at a much higher risk for exposure to HIV and AIDS than the vast majority

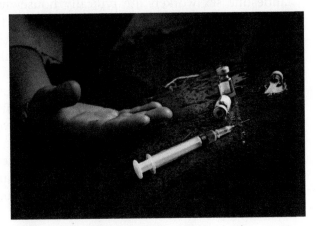

Figure 6-2 Injection drug use is a major route of HIV transmission.

of heterosexuals. While many HIV/AIDS dissidents argue that prostitutes should have a much higher overall prevalence of the infection, the fact is that most of these professional sex workers mandate that their customers use condoms or have other means of barrier protection. The nation of Senegal has learned a hard lesson about the folly of the reasoning that legalized/regulated prostitution will lower the incidence of sexually transmitted infections. Senegal has tolerated prostitution among woman over 21 years of age since 1969. While Senegal has one of Africa's lowest overall

Figure 6-3 Preventing HIV infections among sex trade workers has been proven to be an instrumental part of the fight against AIDS in many countries.

infection rates at less than 1%, this disguises a dangerous rise among vulnerable groups like sex workers. HIV prevalence was not only much higher, but growing among this group. According to Enda Third World, an international NGO based in Dakar, today, HIV infection among legal sex workers in the capital, Dakar, has risen sharply to 21% compared with 1% two decades ago. Another growing problem in areas such as Senegal is the practice of clandestine or survival sex, which is practiced when women are facing an economic dilemma and need money fast. These women typically have no formal education or job skills and are functionally non-literate. Clandestine sex workers are typically housewives, widows, or single women with children. They hide their sexual activities due to shame and stigmatization associated with prostitution and are therefore not registered. While these women do not see themselves as sex workers because they only do this occasionally, their risk of contracting HIV is significantly increased.

Male prostitution has been shown to pose just as high a risk of transmission. The fact is that high rates of HIV have been consistently found among individuals who sell sex in many different countries and cultures. Even in areas where HIV prevalence is generally low, the rate among sex workers is usually higher than the rate found among the general adult population. In some places, sex workers commonly use drugs and share needles and this overlap between sex work and *injecting drug use* is linked to growing HIV epidemics in a number of countries, such as China, Indonesia, Kazakhstan, Ukraine, Uzbekistan, and Vietnam.

HIV Transmission in MSM

MSM are at increased risk for infection with HIV (see Figures 6-1 and 6-4). This is a direct result of the ease with which the virus gets into the bloodstream through infected semen during anal sex. It turns out that rectal tissue is particularly easily disrupted or torn and the trauma of the sexual act leads to this type of tearing. While it is not entirely apparent why there is an increase in unprotected anal intercourse, research evidence suggests several factors, including improvements in HIV treatment leading to more healthy-looking men who are able to have sex without their partners knowing that they are indeed HIV positive.

Seeking sex partners on the Internet is another factor that increases one's chances of ending up with an infected partner. The Internet has the potential to normalize certain high-risk behaviors by making others aware of these behaviors and creating new connections between the men who engage in them.

The use of alcohol and illegal drugs continues to be prevalent among some MSM and is linked to HIV and STI risk. Data show that there is a tendency toward risky sexual behaviors while under the influence of various drugs. These behaviors include

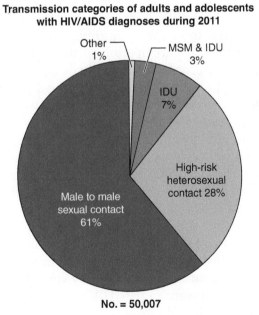

Transmission categories of adults and adolescents with HIV/AIDS diagnoses during 2011

Other 1%
MSM & IDU 3%
IDU 7%
High-risk heterosexual contact 28%
Male to male sexual contact 61%

No. = 50,007

Figure 6-4 Diagram showing HIV transmission categories of all adolescents and adults in the United States *in 2011*. The largest estimated proportion of HIV/AIDS diagnoses among adults and adolescents were men who have sex with men (MSM), followed by persons infected through high-risk heterosexual contact.
(Source: The diagram was created from CDC data. www.cdc.gov/hiv/resources/factsheets/us.htm)

sharing needles or other injection equipment. Maybe most important among these risk factors is a failure to maintain prevention practices such as condom use. The proper and consistent use of latex or polyurethane (a type of plastic) condoms when engaging in sexual intercourse, whether vaginal, anal, or oral, can significantly reduce a person's risk of acquiring or transmitting sexually transmitted diseases, including HIV.

Bug Chasers and HIV Infection

With the devastation that HIV has caused over the past 26 years, it would be logical to conclude that this is the last disease that anyone would want to end up with. It might therefore surprise the average citizen that there are a number of men who are purposely trying to become HIV positive! A **bug chaser** is a gay man who deliberately attempts to contract HIV by having unprotected sex with a man or a group of men who are known to have the virus.

"In private sex clubs across the U.S. men gather for a chance to participate in what is called Russian Roulette. Ten men are invited, nine are HIV-negative, and one is HIV positive. The men have agreed to not speak about AIDS or HIV. They participate in as many unsafe sexual encounters with each other as possible, thus increasing their chances to receive 'the bug.' These are the men known as 'Bug Chasers,'" writes author Daniel Hill (www.alternativesmagazine.com/15/hill.html). The HIV-positive gay man who deliberately transmits the virus to bug chasers is called the "gift giver." Some psychologists theorize that participation in these crazy bug parties is actually an anxiety disorder where the uninfected individuals fear getting HIV so greatly that they would rather contract it and free themselves of the anxiety of living in fear! Bug chasing has been viewed largely with disdain from the gay community and is seen as a self-destructive activity. Leaders of the gay community at large are concerned that the behaviors of bug chasers may contribute to a public perception that the practice is common or encouraged by all gay people and would thus cause further ill will toward them (http://en.wikipedia.org/wiki/Bugchaser). Daniel Hill further reports that there are men who, once infected, feel like they finally "belong"; they are now part of the gay community. Some find relief in knowing that now they don't have to worry about getting infected any more, since the deed is done. Some believe the myth that HIV is a chronic manageable disease and that the new drugs promise them a long and healthy life. Some couples see infection as the deepest level of intimacy. Whatever the reason for this unusual longing to be infected with a deadly virus, I think it is safe to say that most gay men with HIV do not want to pass HIV on, and most gay men who do not have HIV do not want to get infected.

Women and HIV Transmission Risk

Based on data collected to date, gender inequalities as well as biological factors make women and girls especially vulnerable to HIV and to the overall impact of AIDS. Biologically, women are more susceptible to HIV infection than men and are more likely to become infected in any given heterosexual encounter. This is due primarily to the following factors: (1) Women have a greater surface area of mucous membrane exposed during sex compared to their male companions; (2) men transfer a greater quantity of fluids to women during sexual intercourse; (3) male ejaculate fluids contains a higher viral content; and last, (4) the frequency of micro-tears that can occur in vaginal (or rectal) tissue from sexual penetration leave women vulnerable to direct introduction of HIV into the blood circulation.

Gender biases and norms often increase the HIV transmission risk in women and girls. In many areas of the world, cultural norms allow men to have more sexual partners than women and even encourage older men to have sexual relations with much younger women. In certain countries this practice has contributed to higher infection rates among young women (15 to 24 years) compared to young men. Often, women may want to ask their partners to use condoms or to abstain from sex altogether, but are too afraid or lack the power and support to do so. This often leads to men passing HIV on to many partners, often in the same community.

Many women also experience sexual, emotional, and physical violence at some point in their lives. In many of these situations women are forced to have sex, which greatly contributes to HIV transmission due the increase of vaginal tears and lacerations resulting. The threat of violence also prevents women from negotiating safer sexual practices, sharing their HIV status if the results are positive or accessing the necessary treatment. These major challenges are further compounded by the fact that millions of women worldwide do not have access to educational opportunities, resulting in a lack of economic security, which further contributes to their vulnerability to HIV.

Mother-to-Child Transmission   HIV transmission from mother to child during pregnancy, labor and delivery, or breastfeeding is called *mother-to-child* or *perinatal transmission*. There are approximately 1.4 million pregnant women living with HIV in low- and middle-income countries around the world. While HIV testing is routinely performed on all pregnant women in the developed world, only 26% of pregnant women living in these low- and middle-income countries receive HIV tests. In Europe and the United States, about 15 to 20% of babies born to HIV-positive women who are not taking anti-HIV drugs are infected. In 2011, around 330,000 **children** under the age of 15 became infected with HIV and an estimated 230,000 died from AIDS. Almost all of these infections were as a result of mother-to-child transmission and occurred among children living in sub-Saharan Africa While the percentage of HIV-positive women who give birth to HIV-positive infants has declined dramatically in developed countries, due largely to the use of antiretroviral drugs such as **AZT** (Azidothymidine), a lack of prenatal care increases the likelihood that a woman will transmit the virus to her infant. In countries like Uganda, however, mother-to-child transmission is the second most common method of HIV transmission after sexual intercourse. Those women who live in areas where antiretroviral drugs are not readily available transmit the virus to their unborn child or at the time of birth. In sub-Saharan Africa, 40% of all

HIV/AIDS cases result from mother-to-child transmission. Recent evidence confirms that antiretroviral drugs are effective in the prevention of vertical transmission of HIV. The treatment regimen, however, is expensive and many people in developing countries cannot afford the approximately US$980 for both the mother and child to be treated with an AZT regimen. The cost can range from $780 to over $1000 depending on the length of breastfeeding (six months vs. 18 months).

Other Possible Modes of HIV Transmission

There is a low risk of HIV transmission when having a tattoo or body piercing if the instruments contaminated with blood are not sterilized between clients. It is therefore important to determine if people who carry out body piercing or tattoos are following the proper procedures at all times. There are established protocols termed *universal precautions*, which govern the handling of body fluids or instruments that might be contaminated with these fluids, designed to prevent the transmission of blood-borne infections such as HIV and hepatitis B. Under the universal precaution principle, blood and body fluids from all persons should be considered as being infected with HIV, regardless of the known or supposed status of the person. These guidelines may be found on the following World Health Organization (WHO) site: www.who.int/hiv/topics/precautions/universal/en. While there have been unconfirmed stories of HIV transmission at barber shops, this constitutes a very low-risk scenario. Your risk of getting infected with HIV at the barber shop is very low and would only happen if the skin was cut and infected blood got into that wound.

HIV Infection in Healthcare Workers

Healthcare workers may become infected through accidental injuries from needles and other sharp objects that may be contaminated with HIV. The risk is, however, extremely low. It has been estimated that the risk of infection from a needle-stick injury is less than 1%. In the United States, there have only been 56 documented cases of occupational HIV transmissions. All healthcare workers, including dentists, are required to follow the universal precaution guidelines, making the risk of getting HIV from healthcare professionals very low.

While it is true that in the past many people became infected with HIV through blood transfusions and blood products that were contaminated with the virus, these products no longer pose a threat, as whole blood donations and blood products are routinely tested for the presence of HIV-1 and 2. In the early 1980s, many hemophiliacs were accidentally infected with HIV. Hemophiliacs typically lack a critical clotting factor (Factor VIII) needed to prevent severe and possibly fatal bleeding with even a small cut. This factor VIII was made from the batch plasma samples of up to 20,000 pooled human donations. Since no one knew about HIV, it was not assayed for and ended up in many batches of the product that these patients inject directly into their veins daily to survive. There are reports that even after some manufacturers realized that the products might be HIV infected, they did not immediately pull them from the shelves. Many people got infected this way and also infected their spouses (www .thebody.com/whatis/hemophilia.html). Today these products are tightly regulated and are routinely tested for HIV.

You Cannot Get HIV from That!

There are probably more myths about ways in which HIV can be transmitted than those about Bigfoot. While this section will in no way serve as the HIV myth buster narrative, I believe it is important to address a few of the more common ones. While HIV has been found in saliva and tears in very small quantities, contact with saliva, tears, or sweat has never been shown to result in transmission of HIV. The CDC, however, advises against deep French kissing with someone who is HIV positive, since ulcers or abrasions from brushing in the mouth could allow enough viral particles to be transferred. Stories about some bitter HIV-positive person planting syringes of his or her blood and people getting stuck by needles in phone booths, coin returns, movie theater seats, gas pump handles, and other places are highly unlikely. This is because, as stated earlier, the virus is very fragile and unless these needles are attached to syringes with fresh infected blood, they would not be effective in transmitting the live virus.

It is also not possible to get HIV from mosquito bites. When obtaining a blood meal from someone, mosquitoes do not inject blood from any previous individuals that they might have bitten. The mosquito only injects saliva, which acts as a lubricant to enable it to better obtain its meal. Moreover, if mosquitoes could transmit HIV infection, many more young children and preadolescents would have been diagnosed with AIDS, since they are more likely to be exposed to mosquito bites. After much research it appears that HIV is not transmitted by insects. The bottom line is that the virus is not transmitted through normal everyday activities such as shaking hands, toilet seats, swimming pools, sharing cutlery, having contact with animals, regular kissing, sneezes, or coughs. Transmission requires exchange of body fluids from an infected person.

Test Your Knowledge

Name: _____

1. Blood and body fluids can transmit HIV because they contain _____.
   (A) Platelets
   (B) B lymphocytes
   (C) free viruses
   (D) CD4+ cells
   (E) both C and D

2. In a male, the presence of a sexually transmitted diseases (STD) such as gonorrhea or chlamydia can contribute to HIV transmission because ____.
   (A) HIV attaches to these bacteria and are co-transmitted
   (B) STDs can cause erosion of the urethra and provide an entry path for HIV
   (C) the bacterial agents enhance the aggressiveness of HIV
   (D) the body's antibacterial antibodies react with HIV

3. All of the following are methods by which HIV is known to be transmitted except _____.
   (A) anal intercourse
   (B) social kissing
   (C) placental transfer from mother to fetus
   (D) contact with contaminated blood in syringes

4. In which of the following ways can HIV be passed from a mother to her child?
   (A) During pregnancy
   (B) During child birth
   (C) Through breastfeeding
   (D) All of the above

5. True or False: Healthcare workers should assume that all patients with whom they interact are infected with HIV and that the virus is present in their blood and body fluids, using protective wear to prevent transmission.

CHAPTER 7

Prevalence of HIV/AIDS

"In June of 1981 we saw a young gay man with the most devastating immune deficiency we had ever seen. We said, 'We don't know what this is, but we hope we don't ever see another case like it again'" remarked Dr. Samuel Broder, then of the National Cancer Institute. Unfortunately, Dr. Broder's wish was not granted. Since the beginning of the epidemic, more than 60 million people have contracted HIV and approximately 30 million have died of HIV-related causes. Every day, almost 7,000 persons become newly infected with HIV and over 5000 persons die from AIDS, mostly because of inadequate access to HIV prevention and treatment services. Every 12 seconds someone contracts HIV, and every 16 seconds another person dies from AIDS! An estimated 34 million (31.4–35.9 million) people worldwide were living with HIV at the end of 2011 (Table 7-1). An estimated 2.5 million (2.2–2.8 million) became newly infected with HIV and an estimated 1.7 million (1.5–1.9 million) lost their lives to AIDS, despite recent improvements in access to antiretroviral treatment. At the same time, we should not lose sight of the tremendous strides that have been made over the past decade in HIV/AIDS control strategies. There were more than 700,000 fewer new HIV infections globally in 2011 than in 2001. The continent of Africa has cut AIDS-related deaths by one-third in the past six years.

Overall, the HIV incidence rate (the proportion of people who have become infected with HIV) is believed to have peaked in the late 1990s and to have stabilized subsequently, notwithstanding increasing incidence in several countries. Sub-Saharan Africa continues to be the region most affected, with nearly 1 in every 20 adults living with HIV at the end of 2011. Ninety-six percent of people with HIV live in the developing world.

Categories	Estimate	Range
People living with HIV/AIDS in 2011	34 million	31.4–35.9 million
Adults living with HIV/AIDS in 2011	30.7 million	28.2–32.3 million
Women living with HIV/AIDS in 2011	16.7 million	15.4–17.6 million
Children living with HIV/AIDS in 2011	3.3 million	3.1–3.8 million
People newly infected with HIV in 2011	2.5 million	2.2–2.8 million
AIDS deaths in 2011	1.7 million	1.5–1.9 million

Table 7-1 Official global estimates of the HIV/AIDS epidemic as reported by UNAIDS at the end of 2012, and refer to the end of 2011.
(Data Source: http://www.unaids.org/en/resources/presscentre/factsheets/)

Affected Region	Adults and children living with HIV/AIDS	Adults and children newly infected	Deaths of adults and children
Sub-Saharan Africa	23.5 million	1.8 million	1.2 million
Middle East and North Africa	300,000	37,000	23,000
South, South-East, & Eastern Asia	4.83 million	369,000	309,000
Oceania	53,000	29000	1,300
Latin America	1.4 million	83,000	54,000
Caribbean	230,000	13,000	10,000
Eastern Europe & Central Asia	1.4 million	140,000	92,000
Western and & Central Europe	900,000	30,000	7,000
North America	1.4 million	51,000	21,000
Global Total	34.0 million	2.5 million	1.7 million

Table 7-2 Regional statistics for HIV and AIDS end of 2011.
(Data Source: www.unaids.org/en/resources/presscentre/factsheets/)

HIV/AIDS in Sub-Saharan Africa

The region of the world that is most affected by HIV and AIDS is sub-Saharan Africa. Nearly two-thirds of the world's HIV-positive people live in sub-Saharan Africa, although this region contains little more than 10% of the world's population. Heterosexual sex is the main route of HIV transmission here, although certain areas are seeing an upsurge in IV drug use. Most women with HIV in this region have been infected by their husbands or intimate partners. The increasing number of women and children becoming infected with HIV is of great concern to epidemiologists. Women account for 58% of people living with HIV in sub-Saharan Africa. Current data show that in 26 of 31 countries with generalized epidemics, less than 50% of young women have comprehensive and correct knowledge about HIV. Young people, ages 15 to 24 years old, accounted for approximately 42% of all new HIV infections worldwide at the end of 2010. Among young people living with HIV, nearly 80% (4 million) live in sub-Saharan Africa.

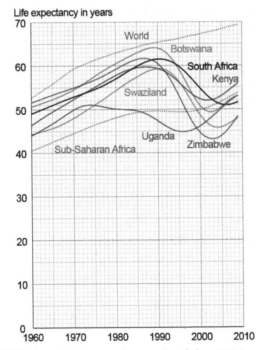

Figure 7-1 Decreased life expectancy at the peak of the AIDS epidemic in some hard-hit African countries.

More than two-thirds (69%) of all people living with HIV (23.5 million), live in sub-Saharan Africa. This number includes 91% of the world's HIV-positive children. In 2011, an estimated 1.8 million people in the region became newly infected. An estimated 1.2 million adults and children died of AIDS, 71% of the world's AIDS deaths in 2011. Sadly, the extent of the AIDS crisis is becoming increasingly clear in many African countries, as the number of HIV-infected people progressing to AIDS continues to rise. It has been estimated that since the beginning of the epidemic, more than 15 million Africans have died from AIDS.

In South Africa, Zambia, and Zimbabwe, young women (aged 15 to 24 years) are three to six times more likely to be infected than young men. This has significantly reduced life expectancy in many countries of the region (Figure 7-1). Ninety-two percent of the world's HIV-positive pregnant women live in sub-Saharan Africa. More than one in five of all pregnant women in six African countries (Botswana, Lesotho, Namibia, South Africa, Swaziland, and Zimbabwe) have HIV/AIDS. In Swaziland, nearly 40% of pregnant women are HIV positive. However, the news is getting better out of these regions. Between 2001 and 2011, in Malawi, the rate of new HIV infections dropped by 73%, in Botswana by 71%, in Namibia by 68%, in Zambia by 58%, and in Zimbabwe by 50%. The country with the largest number of HIV infections, South Africa, reported reduced new HIV infections by 41%. Swaziland, which has the highest HIV prevalence in the world, saw a drop in new HIV infections by 37%.

In sub-Saharan Africa, the estimated number of children under 18 orphaned by AIDS currently stands at 14.8 million. Average life expectancy in sub-Saharan Africa is now 52 years; this is up from 47 years less than 10 years ago, but is still the lowest of any region in the world, which average approximately 70 years. In 2010, Lesotho had sub-Saharan Africa's lowest life expectancy, at 46 years. It is believe that without AIDS, these countries would more closely match the world average.

At the end of 2010, there were 3.5 million children living with HIV around the world. It is estimated that there are 16 million children under 18 years old worldwide who have been orphaned by HIV/AIDS. Over 14.8 million of these children live in sub-Saharan Africa. Gender inequalities as well as biological factors make women and girls especially vulnerable to HIV and to the impact of AIDS.

AIDS is now the leading cause of death in sub-Saharan Africa. Since the beginning of the AIDS epidemic over 33 years ago, more than 25 million Africans have died from AIDS. There is a significant risk that some countries will be locked in a vicious cycle, as the number of people falling ill and subsequently dying from AIDS has a tremendous impact on many parts of African society, including demographic, household, health sector, educational, workplace, and economic aspects. Many grandparents who have lost all of their adult children to the disease are left raising their grandchildren, many of whom also are HIV positive. Clearly, this is a region that continues to need help from the world community and if progress is not made quickly, whole generations will be wiped out by this devastating disease.

AIDS in Eastern Europe and Central Asia

Eastern Europe and Central Asia is the only region where HIV prevalence clearly remains on the rise. The HIV epidemic in Eastern Europe and central Asia continue to grow, with 140,000 people newly infected with HIV in 2011, bringing to about 1.4 million the number of people in the region living with HIV. The number of people living with HIV has almost tripled since 2000. The death toll due to AIDS is also rising sharply. There was a 21% increase in AIDS-related deaths in the region between 2005 and 2011: from 76 000 to 92 000. The HIV epidemics in the region are concentrated mainly among people who inject drugs, sex workers, their sexual partners and, to a much lesser extent, men who have sex with men. The coverage of HIV treatment in the region also remains low with only an estimated 25% of people eligible for HIV treatment currently receiving it. The overwhelming majority of people living with HIV in this region are young. The number of children under 15 living with HIV in Eastern Europe and Central Asia has risen five-fold between 2001 and 2010. The bulk of the people living with HIV in this region can be found in two countries: the Russian Federation and Ukraine, which together account for approximately 90% of infections in the region. The epidemic continues to grow in the Ukraine, with significantly more new HIV diagnoses occurring each year, while the Russian Federation has the largest region of the AIDS epidemic in all of Europe, constituting massive prevention, treatment, and care challenges. Several central Asian republics are experiencing the early stages of the AIDS epidemic, while increasing levels of risky behavior in southeastern Europe suggests that HIV could strengthen its presence there unless prevention efforts are stepped up. According to 2012 country progress reports and UNAIDS estimates, more than 15% of people who inject drugs in Belarus and Tajikistan are living with HIV; more than 20% in Ukraine; and more than 50% in Estonia. Discrimination against vulnerable groups is evident in the Russian Federation, where more than 90% of the people living with HIV were infected through injection drug use, but represent only 13% of those receiving antiretroviral therapy.

HIV/AIDS in Oceania

Oceania collectively encompasses all of Australia, New Zealand, Papua New Guinea, as well as the thousands of coral atolls and volcanic islands of the South Pacific Ocean, including the *Melanesia* and *Polynesia* groups and all the islands of Micronesia. HIV infections remain relatively low in this region where there were an estimated 54,000 people living with HIV at the end of 2011. Over this period 2900 people became newly infected with HIV and an estimated 1300 died of AIDS-related illnesses. **Papua New Guinea** is the country most affected in this region. From 2001 to 2009, there has been 242.857% change in the rate of infection among the population. The majority of reported HIV infections to date have been in rural areas, where more than 85% of the population lives and clustered around concentrations of population, transport routes, and rural enclave enterprises where there are active markets for the exchange and sale of sex. Unsafe heterosexual intercourse is estimated to be the main mode of HIV transmission. Australia and New Zealand have relatively small HIV epidemics with approximately 0.2% of their populations affected. Unprotected sex between MSM accounts for the greatest increase in new cases of HIV infections reflecting a revival of high-risk sexual behavior.

HIV/AIDS in Asia

Overall in Asia, an estimated 4.9 million people were living with HIV in 2011. This number includes the 360,000 people who became newly infected in the past year. Approximately 309,000 died from AIDS-related illnesses in 2011. *These latest estimates are significantly lower than those reported in 2005, when* approximately 8.3 million people were estimated to be living with the disease. This dramatic change in numbers is due mainly to the revised estimates from **China,** and more importantly from **India**, where it was estimated that more than two-thirds of Asia's HIV-infected patients reside. In Cambodia, India, Malaysia, Myanmar, Nepal, Papua New Guinea, and Thailand, the rate of new HIV infections fell by more than 25% between 2001 and 2011. Although national HIV infection levels in Asia are low compared with those of some other continents, notably Africa, the populations of many Asian nations are so large that even low national HIV prevalence means large numbers of people are living with HIV. Like many other regions in the world, risky behavior seems to be the sustaining power for the epidemic. In Asia, there is often more than one form, but at the very heart lies the interplay between injection-drug use and unprotected sex, especially commercial. Cambodia and Thailand have seen an upsurge in HIV infection in MSM, and injection-drug use, while countries like Indonesia and Pakistan will certainly follow along with similar statistics if they do not urgently scale up their responses. In **Thailand**, prevention efforts have resulted in declining levels of HIV since the late 1990s. This decrease was mainly due to fewer men buying sex and the steady increase in condom use. However, recent studies show that premarital sex has become more commonplace among young Thais, and more than one-third of HIV infections in 2009 were among women who had been infected by their long-term partners. Despite the overall achievements in reversing the HIV epidemic in Thailand, prevalence among injecting drug users has remained high over the past 15 years, ranging between 30% and 50%. This continues to be a problem across this entire region. According to 2012 country progress reports, national HIV prevalence among people who inject drugs in Pakistan and Indonesia is more than 25% and 35%, respectively; the reported HIV prevalence in Bangladesh among people who inject drugs was far lower, at less than 2%. HIV prevalence among MSM was more than 15% in Vietnam; 8% in Indonesia; and less than 5% in Bangladesh, Philippines, and Malaysia. This becomes even more troubling in light of new statistics showing that although MSM plays a substantial role in the national HIV epidemics of South and South East Asia, less than 1 in 3 had been tested for HIV in the past 12 months.

The HIV/AIDS epidemic in **China** may be characterized by the following observations: (1) The national prevalence remains low, but the epidemic remains severe in some areas; (2) the number of people living with HIV continues to increase, although new infections have been contained at low level; (3) many patients are now experiencing a gradual progression of HIV to AIDS resulting in an increase of the AIDS-related deaths; (4) sexual contact is the primary mode of transmission, and continues to increase; (5) the epidemic in China is diverse and is currently evolving. There are currently an estimated 780,000 people living with HIV in China, including about 154,000 AIDS patients. Women account for 28.6% of HIV-positive individuals in China. Examination of the routes of transmission among those living with HIV revealed that 46.5% were infected through heterosexual transmission, 17.4% through homosexual transmission, 28.4% through injecting drug use, 6.6% were former blood donors or transfusion recipients, and 1.1% were infected through mother-to-child transmission. As HIV spreads from drug users, sex workers, and their clients to **Asia's** general population, the proportion of HIV infections in women has also been increasing. In 2007, for the first time ever, unsafe sex overtook drug use as the main cause of new HIV infections in China. This confirms fears that the epidemic is moving outward from the high prevalence groups traditionally associated with the HIV epidemic in China. **China's "Four Frees and One Care" program**, which offers free HIV testing, free antiviral drugs to rural residents with AIDS, free drugs to pregnant women, free schooling for AIDS orphans, care and economic assistance to affected households, may provide a model for other nations in supporting families and societies affected by AIDS. China has expanded that program implementing the "Five Expands, Six Strengthens" approach, resulting in important achievements. "Five Expands" means to expand information, education, and communication activities, surveillance and testing, prevent mother-to-child transmission, comprehensive interventions, and coverage of ART. "Six Strengthens" means to strengthen blood safety management, health insurance, care and support, rights protections, organizational leadership, and strengthening of response teams. China has increased the number of people on HIV treatment by nearly 50% in the last year alone.

HIV/AIDS in the Caribbean

Although the Caribbean accounts for a relatively small share of the global epidemic, its HIV prevalence among adults is about 1.0%, which is higher than in all other regions outside sub-Saharan Africa. The number of people living with HIV in the Caribbean remains relatively low with 230,000 in 2011, which has varied little since the late 1990s. The region has seen a sharp decline (42%) in new HIV infections since 2001, from 22,000 in 2001 to 13,000 in 2011. At the same time, the number of AIDS-related deaths fell from an estimated 20,000 in 2005 to 10,000 in 2011. The good news is that the Caribbean region now provides antiretroviral medicines for 79% of pregnant women, approaching near-universal coverage. This is the only region outside of developed nations of the world to do so.

Unprotected heterosexual sex—especially paid sex—is thought to be the main mode of HIV transmission in the Caribbean. Recent reports confirm that HIV prevalence among sex workers is considerably higher than in the general population. In the Dominican Republic, HIV prevalence among sex workers is 4.7%, compared to a national prevalence of 0.7%. According to country progress reports for 2012, HIV prevalence among MSM in Jamaica, Panama, and Chile was 38%, 23%, and 20%, respectively. This is much higher than among the general population in these countries. Haiti, Argentina, and Mexico reported an MSM prevalence of more than 15%. Haiti accounts for more than half (120,000) of people living with HIV in the Caribbean where condom use is very low, despite knowledge of the HIV/AIDS epidemic. Most HIV cases in this region are due to heterosexual contact, with the exception of Puerto Rico and Bermuda, where injection-drug use accounts for a large share of the epidemic. In the Bahamas and Barbados, there are signs that stronger prevention efforts are beginning to help lower HIV infection rates. The HIV epidemics in Jamaica, the Bahamas, and Trinidad and Tobago have also been stable over recent years. The Caribbean's status as the second-most affected region in the world shrouds substantial differences in the islands that are affected. Estimated national adult HIV prevalence is about 1% in Barbados, Dominican Republic, Jamaica, and Suriname; 2% in the Bahamas, Guyana, and Trinidad and Tobago; and exceeds 3% in Haiti. In Cuba, on the other hand, prevalence has not reached 0.2%. The region's epidemics seem to be driven primarily by unprotected heterosexual intercourse, against a common background of poverty, gender inequalities, and a high degree of HIV-related stigma. The ease of migration between islands and countries is a major contributing factor to the spread of HIV while blurring the boundaries between different national epidemics. Commercial sex is a prominent factor in some islands, against a backdrop of severe poverty, high unemployment, and gender-inequality issues. There is no doubt that progress has been made in this region over the past decade; however, there is a severe lack of data on certain high-risk groups indicating gaps in the HIV response of certain Caribbean countries. Stigma and discrimination continue to have a negative impact on the progress that needs to be made in curbing the spread of HIV. Serious prevention initiatives including partnerships with local and regional media groups will go a long way in helping to bring the number of new cases down.

HIV Epidemic in Latin America

The HIV epidemics in Latin America remain generally stable; however, no country in the region has experienced a significant drop in HIV prevalence, and it is projected that the total number of people living with HIV in Latin America will increase in coming years. HIV transmission continues to occur among populations at higher risk of exposure, including sex workers and men who have sex with men. More than half of Latin Americans living with HIV reside in the region's four largest countries: Brazil, Colombia, Mexico, and Argentina. There were a total of 1.4 million people living with HIV in Latin America in 2011. HIV prevalence has, however, decreased by nearly 20% among young people. Significantly, the decline was much higher at 33% among young men, the group where the majority of new HIV infections among young people in the region occur. Overall, the number of people dying from AIDS-related causes declined by 10% between 2005 and 2011 in Latin America. The growth of the epidemic in this region seems to stem from increased rates in Brazil, which accounts for more than one-third of the people living with HIV in Latin America. The serious growth in Brazil stems from MSM, injection-drug users, and people who have heterosexual sex. The light at the end of the tunnel in Brazil is that the country strives to help its people living with HIV; everyone living with HIV has access to antiretroviral drugs through the national health system. The Brazilian government estimates that antiretroviral treatment has contributed to a 50% fall in mortality rates, a 60 to 80% decrease in morbidity rates, and a 70% reduction in hospitalizations among HIV-positive people. In Argentina and Uruguay, HIV remains largely an urban problem. The HIV epidemic is also growing in Venezuela, where the virus is spreading mainly through unsafe heterosexual sex. In Bolivia, infections are mainly among sex workers and MSM. Like many other regions of the world, however, HIV/AIDS in Latin America remains a significant problem. The epidemic in this region is sometimes referred to as a "hidden" crisis. Social stigma, however, has kept many of these epidemics among men who have sex with men hidden and unacknowledged. Awareness remains low, governments have been relatively inactive and surveillance of those affected is sometimes unreliable. There is no doubt that prevention strategies need to be put in place along with mechanisms to tackle discrimination. The good news is that there has been tremendous success in providing antiretroviral drug treatments for those affected in the region. In 2011, coverage of antiretroviral therapy was 68% in Latin America. However, this varies widely from one country to the next. While Argentina, Brazil, Chile, Ecuador, El Salvador, Nicaragua, Paraguay, Peru, and Venezuela achieved more than 60% coverage, in Bolivia, treatment coverage was less than 20% in 2011. It is believed that the global economic crisis of the past decade has had a significant impact on HIV/AIDS prevalence in Latin America.

HIV/AIDS in North America, Western and Central Europe

In North America, Western and Central Europe, the number of people living with HIV has continued to grow, although the number of AIDS deaths remains comparatively low. This is in part because of the availability of antiretroviral therapy coupled with the fact that many of those infected can afford to pay for these drugs. The total number of people living with HIV in North America and Western and Central Europe grew from an estimated 1.8 million in 2001 to 2.3 million in 2011—an increase of 30%. Reports indicate that almost a million people are living with HIV in Western and Central Europe. This number seems to be growing with the resurgence of risky sexual behavior in several countries of the region. Heterosexual contact has emerged as the dominant cause of new HIV infections in several countries in Western Europe. A substantial proportion of new diagnoses are in immigrants originating from countries with serious epidemics, principally countries in sub-Saharan Africa. As a result of antiretroviral drug availability, the number of AIDS deaths plummeted in the late 1990s. HIV prevalence among MSM continues to play a substantial role in national HIV epidemics in these regions. In countries such as France, the Netherlands, and Canada, the HIV prevalence among MSM is reportedly 15% compared to a national HIV prevalence in the general population of less than 0.5% in all three countries. In Western and Central Europe, less than 1 in 3 MSM reported being tested for HIV in the past 12 months, according to 2012 country progress reports. The United States, the Netherlands, and Portugal reported HIV testing coverage among MSM of 50 to 74%, while the reported coverage in Canada was between 25% and 49%. Reported levels of condom use among MSM were less than 50% in the United States, Netherlands, Sweden, and Switzerland. Clearly, this remains a significant problem if the disease prevalence is to be contained.

HIV/AIDS in the United States

HIV incidence peaked in the United States sometime around 1993, then started a steady decline. At the end of 2011, an estimated 1.3 million persons were living with HIV infection in the United States. Of those, 20 percent had undiagnosed HIV infections. The annual numbers of AIDS diagnoses have been relatively constant since 2000; however, new infections continue at far too high of a level, with approximately 50,000 Americans becoming infected with HIV each year. The death rate among people with AIDS has also remained relatively stable in recent years; there were an estimated 20,000 deaths in 2011. Of the new infection in 2011, 77% among males were attributed to male-to-male sexual contact, and 86% among females were attributed to

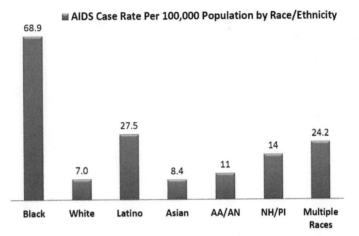

Figure 7-2 AIDS case rate per 100,000 of the population, by race/ethnicity, for U.S. adults/adolescents in 2011. NH/PI is Native Hawaiian/ Pacific Islander; AI/AN represents American Indian and Alaskan Native.

heterosexual contact. Gay, bisexual, and particularly young black/African American MSM, are most seriously affected by HIV. HIV is at pandemic levels in many U.S. cities, especially those in the South, as well as areas with high poverty and a large African American population. Today, nearly half (45%) of Americans report knowing someone living with, or who has died from, HIV/AIDS.

Name: _____

1. Which region of the world is most affected by the HIV epidemic?
   (A) The Caribbean
   (B) Southeast Asia
   (C) Western Europe
   (D) sub-Saharan Africa
   (E) Latin America

2. _____ is the main mode of HIV transmission in the Caribbean.
   (A) Men who have sex with men
   (B) Injection drug use
   (C) Unprotected heterosexual sex
   (D) Mother-to-child transmission
   (E) Both A and B

3. Which of the following represents the group at greatest risk for HIV transmission in the United States?
   (A) Men who have sex with men
   (B) Injection drug users
   (C) High-risk heterosexual males
   (D) Heterosexual females
   (E) Children of HIV-infected mothers

4. True or False: Since the beginning of the AIDS epidemic, more than 60 million people have contracted HIV and approximately 30 million have died of HIV-related causes.

5. Which country has the "Four Frees and One Care" as well as the "Five Expands, Six Strengthens" programs to combat HIV/AIDS?

CHAPTER 8

Race, Poverty, and HIV/AIDS in the United States

HIV/AIDS and Race

There is no doubt that there have been significant successes in addressing HIV/AIDS around the world, including in the United States; however, many challenges remain. Today, there are more people living with HIV than ever before. Millions of new infections occur each year, and since the beginning of the HIV and AIDS epidemic, well over half a million people have died of AIDS in the United States. Approximately 50,000 people become newly infected each year, and many are not aware of their HIV status. Blacks and Latinos account for a disproportionate share of new HIV infections, relative to their size in the U.S. population. Hispanics/Latinos represented 16% of the population but accounted for 21% of new HIV infections and 19% of people living with HIV infection in 2010. Infections for Latino males was 2.9 times that for white males, and the rate of new infections for Latinas was 4.2 times that for white females.

Blacks account for more new HIV infections, people estimated to be living with HIV disease, and HIV-related deaths than any other racial/ethnic group in the United States. Currently, the HIV epidemic among African Americans is a major health crisis. The epidemic has also had a disproportionate impact on black women, youth, and men who have sex with men, and its impact varies across the country. At the end of 2011, there were approximately 1.3 million people living with HIV/AIDS in the United States, including more than 510,000 who are black. Blacks make up approximately 12% of the U.S. population. However, in 2011, blacks accounted for 46% of all new cases of HIV/AIDS diagnoses in the United States (Figure 8-1). There is no doubt that several factors contribute to the continued HIV epidemic among African Americans, including poverty, lack of access to good healthcare, higher rates of some sexually transmitted infections, lack of awareness of HIV status, and continued stigma. The primary transmission category for black men living with HIV/AIDS was sexual contact with other men, followed by injection-drug use and high-risk heterosexual contact. For black women living with HIV/AIDS, the primary transmission category is high-risk heterosexual contact, followed by injection-drug use.

Since the epidemic began, more than 260,800 blacks with an AIDS diagnosis have died. HIV was the fourth leading cause of death for black men and black women, ages 25–44 in 2010, ranking higher than any other racial or ethnic group in the United States. This rate of AIDS diagnosis among blacks has risen significantly from 25% of all cases in 1985 to 49% at the end of 2011. The estimated rate of new HIV infections for black women (38.1/100,000 population) was 20 times as high as the rate for white women, and almost five times as high as that of Latinas. The AIDS case rate per 100,000 among African American adults and adolescents was more than nine times that of whites in 2011 (Figure 7.3). The AIDS case rate for black men (79%) was the highest of any group, followed by black women (39%). By comparison, the rate among white men was 11.2. While both white and black American males are

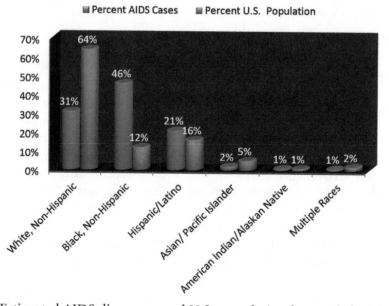

Figure 8-1 Estimated AIDS diagnoses and U.S. population by race/ethnicity. Note that African Americans, who make up 12% of the U.S. population, account for almost half of all new HIV infections in 2011.

(*Source: This figure was generated from data distributed by the Keiser Family Foundation*)

more likely to have been infected through sex with other men, heterosexual trans-mission and injection drug use account for a greater share of infections among black men than white men. A recent study in five major U.S. cities found that 39% of black MSM in the study were infected with HIV, compared to 34% of white MSM and 23% of Latino MSM. At some point in their lifetimes, an estimated 1 in 16 black men and 1 in 32 black women will be diagnosed with HIV infection. With the greater number of people living with HIV in African American communities and the fact that African Americans tend to have sex with partners of the same race/ethnicity, this community face a greater risk of HIV infection with each new sexual encounter. This disturbing trend of HIV infection among African Americans in general appears to be influenced by several factors. These include but are not limited to (1) lack of access to and use of healthcare system; (2) lack of health insurance; and (3) geography (10 states account for 71% of all HIV cases among blacks). It is clear that something needs to be done immediately to address this growing epidemic among African Americans. If this is not done very soon, the HIV/AIDS epidemic will again become a very serious healthcare crisis in the United States while everyone is donating resources to and focusing on the problem in Africa and Asia.

HIV/AIDS in U.S. Poverty Areas

An HIV/AIDS epidemic is defined by the percentage of the population living with HIV. An epidemic is considered to be generalized when the HIV prevalence is 1% or more in the general population. It is said to be concentrated when the HIV prevalence is below 1% in the general population but exceeds 5% in specific at-risk populations such as IV drug users or sex workers. The epidemic is said to be low level if the HIV prevalence is not recorded at a significant level in any group. At the XVIII International AIDS Conference held in Vienna, Austria, in 2010, the CDC reported the results of a study it had recently conducted on the relationship between poverty and HIV/AIDS prevalence. The data revealed that there was a higher HIV infection rate in U.S. poverty areas and that the link to poverty was more significant than links to race. The data further confirmed that the local HIV/AIDS epidemics in some large U.S. cities have reached the status of generalized epidemic. The CDC concluded that poverty, rather than race/ethnicity, was the major demographic factor influencing HIV prevalence among heterosexuals in economically disadvantaged urban areas. The data revealed that 2.1% of heterosexuals living in high-poverty urban areas in the United States are infected with HIV. The study, which included over 9,000 individuals across 23 U.S. cit-ies defined as high-poverty areas according to the U.S. Census Bureau, included areas in which at least 20% of residents have household incomes below the poverty line. In November 2012 the U.S. Census Bureau said more than 16% of the population lived in poverty in the United States. The Census Bureau uses a set of money income thresholds that vary by family size and composition to determine who is in poverty.

The poverty level for 2012 was set at $11,170 for an individual or $23,050 (total yearly income) for a family of four. The HIV prevalence rates in urban poverty areas were inversely related to annual household income (Figure 8-2).

Nationally, the United States is considered to have a concentrated HIV epidemic, where higher rates of infection are confined mainly to individuals who engage in high-risk behaviors, which in the United States are primarily gay and bisexual men and injection drug users. However, within the low-income urban areas included in the study, individuals living below the poverty line were at greater risk for HIV than those living above it (2.4% prevalence vs. 1.2%). The prevalence for both urban groups was far higher than the national average, which currently stands at 0.45%. In contrast to the racial disparity for HIV infection seen across the United States, there were no significant differences in HIV prevalence by race or ethnicity in these low-income urban areas. Prevalence among blacks was 2.1%, among Hispanics it was also 2.1%, and 1.7% among whites. The general U.S. prevalence rate for blacks is almost eight times that of whites, and the prevalence rate among Hispanics is nearly three times that of whites (Figure 8-3).

The nation's capital, Washington, DC, is not only a place of significant poverty, low incomes, and high unemployment. It also holds the distinction of having the highest HIV infection rate in the United States. It has been determined that if the entire population of Washington, DC was an African nation, it world rank 23rd out of 54 in the world in terms of HIV prevalence. DC has an HIV positive rate of 3.2%. This is higher than that of Ethiopia (1.4%), Rwanda (2.9%), D.R. Congo (1.3%), and

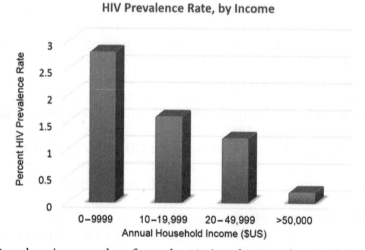

Figure 8-2 Based on income data from the National HIV Behavioral Surveillance System and heterosexuals at risk for HIV infection (NHBS-HET), the lower the income level, the greater the HIV prevalence rate.
(Data source: www.cdc.gov/hiv/risk/other/poverty.html).

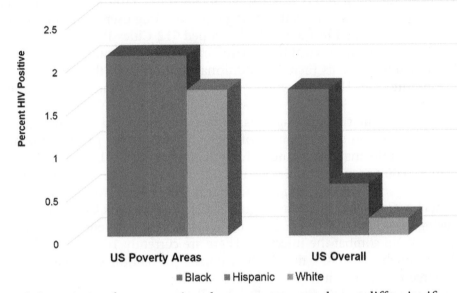

Figure 8-3 HIV prevalence rates in urban poverty areas do not differ significantly by race or ethnicity. This is in contrast to the substantial racial and ethnic differences found in rates for the overall United States.

Sierra Leone (1.5%), all of which receive PEPFAR funds to combat their epidemic. The U.S. President's Emergency Plan for AIDS Relief (PEPFAR) is the U.S. government initiative to help save the lives of those suffering from HIV/AIDS around the world. The epidemic in the District of Columbia is driven by a complex interplay of factors. Transmission patterns vary by race/ethnicity, where blacks are more likely to be infected through heterosexual contact, whereas whites and Latinos are more likely to be infected through sex between men. The epidemic is also not uniformly distributed in the city and can vary widely from one Ward to the next. In seven of DC's eight wards (all but Ward 3), more than 1% of adult/adolescent residents are living with HIV. Prevalence ranges widely, from a high of 3.1% in Ward 8 to a low of 0.5% in Ward 3. The disturbing news from a survey conducted in 2007 shows that while over 61% of those living with HIV in DC know their status, 53% of men and 40% of women reported having overlapping sex partners in the last year, and only 30% reported using a condom the last time they had sex. The vast majority of people living in the communities hardest hit by HIV/AIDS in DC live at or below the poverty line, lack proper healthcare and housing, and to a great extent feel that they have been abandoned by government officials, civic leaders, and their country on whole, which sends millions to fight the epidemic in faraway places with long-term commitments, while many perish while looking at the symbol of the seat of power in the free world out their windows. The fact is that government officials have started to listen and efforts have been scaled up in recent years, producing measurable successes, including decreases in new diagnoses, especially among injection drug users and in infants. DC is one of the areas included in the newly developed "12 Cities Project," an initiative under the U.S. Department of Health and Human Services targeting areas hard hit by HIV in the United States. This includes additional support from the CDC and better surveillance monitoring. The other good news in this crisis in the nation's capital is that the vast majority (89%) of those newly diagnosed with HIV are linked to care within 12 months of their initial diagnosis. In fact, it is estimated that currently 76% are linked to care within three months of their diagnosis. HIV remains mainly an urban disease, with the majority of individuals diagnosed with AIDS in 2010 residing in areas with 500,000 or more people. Areas hardest hit (by ranking of AIDS cases per 100,000 people) include Baton Rouge, LA; Miami, FL; Jackson, MS; Baltimore, MD; New Orleans, LA; Columbia, SC; Washington, DC; and New York City, NY. The CDC has been using the criteria of areas hardest hit according to prevalence rate to allocate much need funds to combat the infection. There are currently 10 major cities receiving funding through the CDC's new highest-impact prevention strategies using this geographic targeting approach (Figure 8-4).

According to a recent CDC estimate, 50% of all new AIDS diagnoses in the United States are in the following states: Alabama, Arkansas, Florida, Georgia, Louisiana, Mississippi, North Carolina, South Carolina, and Tennessee. The South has the largest numbers of adolescents and adults living with HIV, the highest rate of AIDS deaths of any U.S. region, and the fewest resources to fight the epidemic. HIV/AIDS in the South is concentrated largely in poor minority-dominated communities where diagnoses tend to occur late in the infection, often only after the infection has progressed to AIDS. While thousands wait to get medications, many of those who receive treatment do not respond well because of the late stage of their disease. In Mississippi,

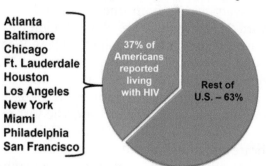

Figure 8-4 Ten metropolitan cities eligible for direct CDC funding. These 10 cities account for approximately 37% of all Americans living with an HIV diagnosis.

for example, the AIDS mortality rate is 60% higher than the national average, and approximately 50% of people who know they are HIV positive are not receiving care. More than half of all poor and black households are in the South. The poverty rate is higher among African Americans (28%) than for any other race. The socioeconomic issues associated with poverty—including limited access to high-quality healthcare, housing, and HIV prevention education—directly and indirectly increase the risk for HIV infection and affect the health of people living with and at risk for HIV infection. Poverty is also associated with higher rates of incarceration, which affects gender balance, increases partnership concurrency, and disrupts relationships. These combined factors lead to increase HIV infection rates.

Among the 10 states with the lowest high school graduation rates (<65%), seven are in the southern United States. According to the U.S. Census Bureau, rates are rising faster in the southern United States than in other region. There has been on average an annual increase of approximately 1.2% in poverty in the South, about double the rise seen in the Northeast, Midwest, or West. The share of the U.S. population earning below half of the federal poverty line has risen to 6.7%, a record high. According to a study by U.S. Department of Agriculture, 9 out of every 10 black Americans who reach the age of 75 spend at least one of their adult years in poverty. It has been estimated that by the age of 25, 48.1% of African Americans will have experienced at least one year in poverty. By age 40, the number grows to 66% and to more than 75% by age 50. More than 90% will have lived below the poverty line by age 75. Researchers say that by age 28, the black population will have reached the cumulative level of lifetime poverty that the white population arrives at by age 75. Blacks have experienced in nine years the same risk of poverty that whites experience in 56 years. It is therefore clear that poverty matter a great deal in the HIV/AIDS epidemic. However, poverty cannot be separated from race. "The fact that almost every black American will experience poverty at some point during their adulthood speaks volumes about AIDS in America. Poor people get AIDS. Black people are poor," says Phill Wilson, President and CEO of the Black AIDS Institute. Stigma, fear, discrimination, and negative perceptions about HIV testing can also place African Americans at higher risk. Many at risk for infection fear stigma more than infection and may choose to hide their high-risk behavior, rather than seek counseling and testing. While the general epidemic in the United States peaked sometime in the mid-1990s, the rate of the infection in the black community as well as urban poverty areas continues to grow. Now, the potential impact has been recognized by the U.S. federal government and the CDC, and its partners are pursuing a high-impact prevention approach to advance the goals of the National HIV/AIDS Strategy and maximize the effectiveness of current HIV prevention methods, especially in the worst affected areas.

Test Your Knowledge

Name: _____

1. Which racial/ethnic group accounts for the greatest number of new HIV infections in the United States?
   (A) Whites
   (B) Blacks
   (C) Latinos
   (D) Pacific Islanders
   (E) Multiracial

2. In order for an epidemic to be considered generalized the prevalence has to be greater than _____ in the general population.
   (A) 0.5%
   (B) 1%
   (C) 5%
   (D) 10%
   (E) 25%

3. True or False: U.S. urban poverty areas have an HIV prevalence rate that is higher than the general population.

4. True or False: Poverty is a greater factor than race in HIV prevalence in U.S. urban cities.

5. True or False: The northeastern region of the United States has almost three times the new cases of HIV and AIDS deaths than the southern states.

6. True or False: At some point in their lifetimes, an estimated 1 in 16 black men and 1 in 32 black women will be diagnosed with HIV infection.

7. True or False: The poverty rate is higher among African Americans than for any other race.

8. Which U.S. city has the highest prevalence rate of HIV?

CHAPTER 9

Testing and Diagnosis of HIV/AIDS

The first HIV tests developed in 1985 were designed to screen blood for transfusion purposes. Back then, there were no treatments available for HIV infection, and it was not even known if an HIV-infected person would get AIDS or how quickly it might happen. It seemed at the time that there was no pressing reason for people to find out their HIV status. There were people, however, who, based on their activities, suspected that they might be infected and wanted to know their status. It was feared that these individuals were going to donate blood just to get tested. That soon changed with the advent of antiretroviral therapy and prophylaxis (preventive treatment) aimed at preventing opportunistic infections. Over time, HIV testing became a gateway to treatment, as well as a prevention tool. There are specific tests available that can readily determine whether a person is infected with HIV.

Although there have been many technological advances since 1985 where testing is concerned, HIV testing still follows the same basic testing procedure as in 1985: HIV infection is only considered confirmed after two tests have been done—a screening test and a confirmatory test. These tests look for antibodies to HIV or HIV antigens in the patient's blood or other body fluids. These antibodies are proteins produced by the immune system to fight an infection. Most antibody tests are used for diagnostic purposes to determine if someone is infected. There are, however, additional HIV tests that are used to monitor the status of a person who already knows that he or she is infected with HIV. These help measure how quickly the virus is multiplying (viral load test) or the health of your immune system (CD4 cell count). HIV testing is integral to prevention, treatment, and care efforts. Knowledge of one's HIV status is important for preventing the spread of disease. Test results can also reduce stress and undue anxiety if the result is negative. If a person is HIV positive, testing provides an opportunity to receive risk reduction counseling and other useful information. Many who learn they are HIV positive modify their behavior to reduce the risk of HIV transmission. Early knowledge of HIV status is also important for linking those who are HIV positive to medical care and services that can reduce morbidity and mortality and improve quality of life.

HIV antibody tests do not work as soon as an individual becomes infected. This is because it usually takes several weeks to generate antibodies to HIV. The period from the time of infection to the point where antibodies are detected is called the *window period*. Most people generate a detectable immune response within four to six weeks, although approximately 5% of infected individuals take up to three months. Therefore, taking an antibody test less than four weeks after exposure will not tell you very much. If one tests negative after four weeks and was involved in a high-risk encounter, he or she might be asked to return three weeks later to do a second test, just to be certain (Figure 9-1). In the UK and Ireland, most clinics also perform the Recent Infection Testing Algorithm (RITA) test. RITA is a generic name for a number of laboratory techniques that distinguish recent and established HIV infection and is primarily used for monitoring at a population level. The principle of RITA-based monitoring is predicated on the fact that production of HIV-specific immunoglobulin M (IgM) normally peaks one to two weeks after infection. IgM then falls to background levels while levels of immunoglobulin G (IgG) increase slowly over several months and are maintained during chronic infection. RITA monitors the levels of HIV-specific IgM and IgG in patient sera.

HIV Testing for Pregnant Women

Approximately 25% of HIV-infected pregnant women who are not treated during pregnancy can transmit HIV to their infants during pregnancy, during labor and delivery, or through breastfeeding. In 1998, the Institute of Medicine (IOM) published a report that recommended simple, routine, and voluntary HIV testing for all pregnant women in antenatal settings, given the effective interventions available to treat HIV-infected women and reduce risk for perinatal HIV transmission. This HIV test should be offered to all pregnant women as part of the standard battery of prenatal tests, regardless of risk factors and the prevalence rates in the community. The IOM report clearly stated that whenever a woman of childbearing age is unaware of her HIV status or her risk for HIV or when an HIV-infected pregnant woman does not know her status, an opportunity is missed. This means that the woman (a) does not receive prenatal care, (b) is not offered HIV testing, (c) is unable to obtain HIV testing, (d) is not offered chemoprophylaxis, (e) is unable to obtain chemoprophylaxis, or (f) does not complete the chemoprophylaxis regimen. Prophylaxis failures occur when an infant becomes

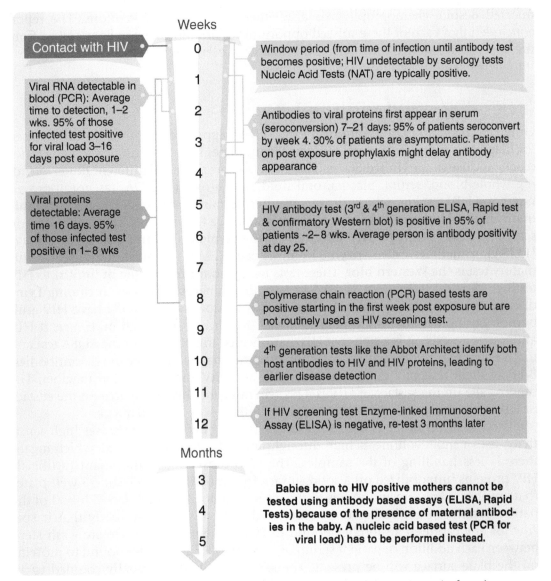

Weeks

Contact with HIV

0 — Window period (from time of infection until antibody test becomes positive; HIV undetectable by serology tests Nucleic Acid Tests (NAT) are typically positive.

1

Viral RNA detectable in blood (PCR): Average time to detection, 1–2 wks. 95% of those infected test positive for viral load 3–16 days post exposure

2 — Antibodies to viral proteins first appear in serum (seroconversion) 7–21 days: 95% of patients seroconvert by week 4. 30% of patients are asymptomatic. Patients on post exposure prophylaxis might delay antibody appearance

3

4

Viral proteins detectable: Average time 16 days. 95% of those infected test positive in 1–8 wks

5 — HIV antibody test (3rd & 4th generation ELISA, Rapid test & confirmatory Western blot) is positive in 95% of patients ~2–8 wks. Average person is antibody positivity at day 25.

6

7

8 — Polymerase chain reaction (PCR) based tests are positive starting in the first week post exposure but are not routinely used as HIV screening test.

9

10 — 4th generation tests like the Abbot Architect identify both host antibodies to HIV and HIV proteins, leading to earlier disease detection

11

12 — If HIV screening test Enzyme-linked Immunosorbent Assay (ELISA) is negative, re-test 3 months later

Months

3

4 — **Babies born to HIV positive mothers cannot be tested using antibody based assays (ELISA, Rapid Tests) because of the presence of maternal antibodies in the baby. A nucleic acid based test (PCR for viral load) has to be performed instead.**

5

Figure 9-1 Time from sexual exposure until HIV antibodies against viral antigens can be detected.

infected despite chemoprophylaxis and other preventive interventions. The report concluded that each of these missed opportunities or failures deserves attention from the appropriate service providers and available prevention programs (www. cdc.gov/ mmwr/preview/mmwrhtml/rr5019a2.htm).

Screening and Confirmatory Tests for HIV

Only U.S. Food and Drug Administration (FDA)–approved HIV tests should be used for diagnostic purposes. Several HIV test technologies have been approved by FDA for diagnostic use in the United States. These tests enable testing of different fluids including whole blood, serum, plasma, oral fluid, and urine. The FDA has not approved home-use HIV test kits, which allow consumers to purchase a test kit, collect a sample in private, and interpret their own HIV test results in a few minutes. The most common screening test available for HIV infection is the **enzyme-linked immunosorbent assay (ELISA),** sometimes called enzyme immunoassay (EIA). The most often used confirmatory test is the **Western blot**. These tests are by no means unique or limited to HIV testing. Identical technology is used in tests for numerous illnesses, including Lyme disease, hepatitis, and cytomegalovirus. The body does not naturally have HIV antibodies. Antibodies are only produced in the presence of an infection. Hence, if HIV antibodies do exist, it is a sign that HIV has invaded the body. Both the ELISA test and the Western blot test detect HIV infection by detecting the presence of HIV antibodies. Both of these tests rely on the fact that HIV and HIV antibodies bind together. The binding site on the surface of HIV is a protein antigen. There is an area on the surface of HIV antibodies into which HIV antigens fit, like a key fits into a lock.

The ELISA technique is used for initial screening because it possesses high sensitivity, uses a small volume, is high throughput, and is fully automated, which means there is less handling of the samples. The ELISA test uses recombinant (artificial) HIV proteins immobilized on a solid surface (typically inside a plastic 96-well plate) that are able to capture antibodies to the virus contained in the infected blood of the patient. Captured antibodies can be detected using a second antibody that is specific for any captured antibodies bound to HIV proteins. Since there are wash steps between each addition of patient serum or reagent, only antibodies bound to proteins on the plate surface will be present. The second antibody is usually coupled to an enzyme (hence, the name) that when introduced to its specific substrate (molecules upon which an enzyme acts) produces a colored product (Figure 9-2). The change in color is read by a machine that determines the optical density based on the intensity of the color. The intensity of this color is directly proportional to the amount of HIV-specific antibodies in the patient's blood. This test can be performed on either blood or urine. If the ELISA test is positive, then the confirmatory Western blot test is done. The Western blot is also an antibody-based test; however, this test is more specific than the ELISA.

1. Blood sample is drawn
   from person to be tested

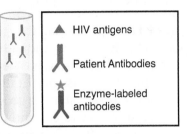

▲ HIV antigens

Ⅰ Patient Antibodies

Ⅰ Enzyme-labeled
  antibodies

If person is infected with HIV,
there will be antibodies in the serum

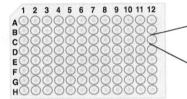

2. Microplate wells are coated with
   HIV proteins

3. The patient serum is added and
   antibodies bind to HIV proteins

4. A second enzyme-labeled antibody
   that will only bind to the first
   antibody is added

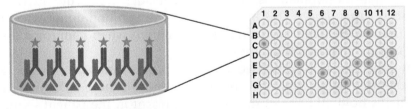

5. A substrate is added that reacts with the enzyme, to form a colored product, indicating
   that patient antibodies are present.

Figure 9-2 Diagram of an ELISA to detect HIV-specific antibodies. The intensity of the color in each well is directly proportional to the amount of HIV antibodies in the patient's blood.

Fourth-Generation HIV Tests

Early identification of HIV disease has the potential to significantly reduce disease transmission. Approximately 21% of individuals with HIV infection in the United States are unaware of their status, and it has been estimated that these persons account for more than 50% of new transmissions of the disease. First-generation ELISA testing can detect HIV antibodies in sera starting around five to six weeks. Third-generation tests are the ELISA screening assays currently used, and they are able to detect HIV antibodies at four weeks post-infection in most patients. The most sensitive third-generation antibody tests will pick up about 42% of acute-phase infections. Fourth-generation HIV tests are ELISA-based assays that have been successfully used for the past decade in many parts of Europe. The FDA approved the first fourth-generation assay for use in the United States in June 2010. Fourth-generation assays have greater sensitivity because of their ability to detect the p24 antigen as well as conventional HIV-specific antibodies in the serum. Identification of highly contagious individuals via p24 antigen detection during the acute phase of infection affords the opportunity of reducing the viral load through antiretroviral medication, as well as initiating behavioral modification. With the current third-generation assays, HIV antibodies can be detected in most individuals within three to four weeks of viral transmission. However, the p24 antigen level is elevated and becomes detectable before an antibody response during the initial phase of viral replication. Currently, the only FDA-approved fourth-generation assay is the Abbott Architect HIV Ag/Ab Combo Assay. The assay features qualitative detection of HIV p24 antigen and antibodies to both HIV-1 and HIV-2 in serum and plasma. Compared with a conventional FDA-approved HIV-1/2 antibody assay in seroconversion testing, the Architect HIV Ag/Ab Combo is able to detect HIV infection approximately a week earlier, significantly reducing the diagnostic window period. It reportedly detects 95% of all HIV infections within four weeks. Fourth-generation HIV testing is indicated for both routine and perinatal testing but is most beneficial in high-risk populations in whom the incidence of acute infection is high. The Architect is currently not FDA approved for routing blood donor screening; however, it has FDA endorsement for use in urgent situations when traditional donor HIV screening assays are unavailable or impractical.

HIV, like any other virus, is composed of a number of different proteins. HIV positivity can therefore only be confirmed by the presence of these proteins. If HIV antibodies to proteins from all parts of the virus are present in the patient's sample, this confirms seropositivity, and there is very little doubt that the individual is infected with the virus. The antibodies seen in a positive sample are typically reactive with all classes of HIV gene products. Table 9-1 lists the proteins most likely to react with patient antibodies from infected sera on a Western blot.

For the Western blot, the recombinant HIV proteins are separated by electrophoresis on a sodium dodecyl sulfate polyacrylamide gel electrophoresis (SDS-PAGE)

Protein	Origin and Function
gp160	viral envelope precursor (env)
gp120	viral envelope protein (env) binds to CD4
p24	viral core protein (gag)
p31	Reverse Transcriptase (pol)

Table 9-1 HIV proteins typically seen reacting with patient antibodies on a Western blot. Please note that there might be more proteins than these on any positive blot. Gene products from each area of the virus are represented here (envelope, capsid, and core proteins).

system based on their individual molecular weight. The larger proteins are at the top of the gel matrix, and the smaller proteins that travel faster are toward the bottom. Once these proteins have been separated on the gel, they are transferred by electrical charge to a solid paper-based membrane. This membrane is usually made from nitrocellulose or polyvinylidene fluoride (PVDF). The proteins will be transferred in the same pattern as seen on the gel. The membrane is then cut into strips to facilitate testing of a large number of samples for antibodies directed against the blotted protein (antigen). The patient's serum sample is then taken and added to the paper membrane with the HIV proteins on it and allowed to interact. If the patient is HIV positive, there will be antibodies made against various parts of the virus floating around in the serum. Each antibody will react with the specific protein against which it was made on the blot membrane. Since these proteins (antigen and antibodies) are so small, you will not be able to see this reaction with the naked eye or even with a microscope; therefore, some type of "label" must be placed on this reacting complex to visualize the reaction. When a labeled secondary antibody is added in a similar manner as the ELISA test above, it will bind to the patient antibodies that are bound to the membrane. A substrate will then be added and a colored product will be formed, resulting in a dark precipitate at the sites where there are bound antibodies on the blot.

Interpretation of Western Blot Data

In 1997 the Centers for Disease Control along with several other organizations established criteria for serologic interpretation of HIV Western blot tests (see Figure 9-3.). While different countries have their own standards for what constitutes a positive HIV Western blot, the CDC and FDA have arrived at a decision that must include a specific combination of HIV proteins. If no viral bands are detected, the result is negative (Table 9-2). If at least one viral band for each of the *gag, pol,* and *env* gene-product groups is present, the result is positive (see Table 9-1). Western blot tests in which fewer than the required number of viral bands are detected will be reported as

Banding Patterns Observed	Reported Results
No bands present	Negative
Bands at either p31 OR p24 AND bands present at either gp160 OR gp120	Positive
Bands present, but pattern does not meet criteria for positivity	Indeterminate

Table 9-2 Table of banding patterns on a Western blot and interpretations that would be reported.

indeterminate. A person who has an indeterminate result should be retested, since later tests may be more conclusive.

Research data show that almost all HIV-infected persons with indeterminate Western blot results will develop a positive result when tested one month after the first test. If an individual remains persistently indeterminate over a period of six months, this would suggest that the results are not due to HIV infection and therefore are the result of a false-positive reaction. Indeterminate Western blot results can be caused by either incomplete antibody response to HIV in samples from infected persons or nonspecific reactions in samples from uninfected persons.

Incomplete antibody responses that produce negative or indeterminate results on Western blot tests can occur among persons recently infected with HIV who have low levels of detectable antibodies (i.e., seroconversion), persons who have end-stage HIV disease, and perinatally exposed but uninfected infants who are seroreverting (i.e., losing maternal antibody). Nonspecific reactions producing indeterminate results in uninfected persons seem to occur more frequently among pregnant or parous women than among other persons. False-positive Western blot results (especially those with a majority of bands) are rare.

False positive tests results using the ELISA tests will require secondary testing using this Western blot procedure. Western blot results help confirm if someone is HIV positive because some conditions (including lupus and Lyme disease) may yield a false-positive ELISA test result. The blots of real patient data along with positive and negative controls shown in Figure 9-4 are very typical of what is seen in the clinical setting. While it is clear that sample number 3 is negative and that numbers 2 and 4 are clear positives, sample 1 does not have either a gp120 or a gp160 band to represent the envelope proteins. It therefore is an indeterminate result, and the patient in this case would be asked to repeat the test in approximately one month. An HIV test should be considered positive only after screening (ELISA) and confirmatory tests (Western blot) are reactive. A confirmed positive test result indicates that a person has been infected with HIV. Please note, however, that it does not mean that the person has

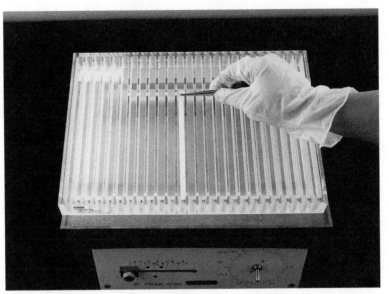

Figure 9-3 A technician is placing a Western blot strip in a test tray containing patient serum. Antigens have already been transferred from the gel onto the test strip. Antibodies, if present in the patient serum, will bind to this sheet and later be detected with a labeled secondary antibody.
(Source: CDC, http://phil.cdc.gov/phil_images/20021205/23/PHIL_2613.tif)

AIDS. Incorrect HIV test results occur primarily because of specimen-handling errors, laboratory errors, or failure to follow the recommended testing algorithm.

Nucleic Acid–Based HIV Testing

Both the ELISA and Western blot tests are indirect assays, meaning they do not detect HIV. Instead, they detect the immune system response to HIV; in the absence of a developed immune system, they would be ineffective. This would be the case if the HIV status of an infant needed to be determined. Furthermore, neither the Western blot nor the ELISA protocols help the physician determine the viral load. As discussed in the previous section dealing with the progression to AIDS, determination of the level of HIV in the peripheral blood is very important in establishing disease stage and treatment options, as well as the effectiveness of a treatment regimen.

The most effective tests that directly assay for the presence of the HIV particles involve the use of the very versatile **polymerase chain reaction (PCR)**. The PCR chemically multiplies viral DNA that exists in the sample by a factor of approximately one million (this is the "chain reaction," which its name describes). The PCR reaction can detect HIV DNA or RNA in the patient's blood. The test is designed to detect

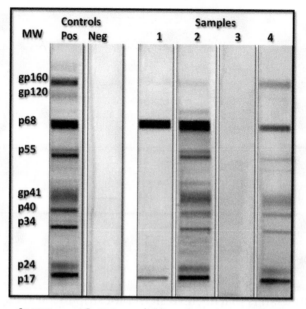

Figure 9-4 Picture of HIV-specific Western blot showing controls and patient samples. Sample number 3 is negative, while samples 2 and 4 are strong positives, based on the CDC criteria. Sample 1 falls into the category of indeterminate result and would be retested before the results can be confirmed. Although there are two strong bands present, they are both from internal HIV proteins and there are no surface glycolipids present.

and amplify a specific fragment of the HIV DNA in a relatively short period of time (Figure 9-5).

Since the viral RNA is first converted into double-stranded DNA before being integrated into the host cell DNA or genome, PCR may be used to detect this proviral DNA. For this type of test, blood is drawn from the patient's arm; the peripheral blood mononuclear cells (PBMCs), which are circulating white blood cells, are isolated. DNA is then isolated from the cells and PCR amplification of a specific short fragment is accomplished using a machine called a Thermocycler. This amplified DNA is then quantified to determine the viral load. There are PCR protocols available that can detect 1 cell containing proviral DNA in a sample of 150,000 cells. HIV infection can be reasonably excluded in an infant if PCR tests performed at one and four months are negative for HIV DNA. Definitive exclusion of infection does not take place, however, until HIV antibody testing is negative. Newborns do not have a developed immune system and would therefore not have HIV-specific antibodies in their blood. Moreover, newborns receive antibodies from their mothers through the placenta. The HIV

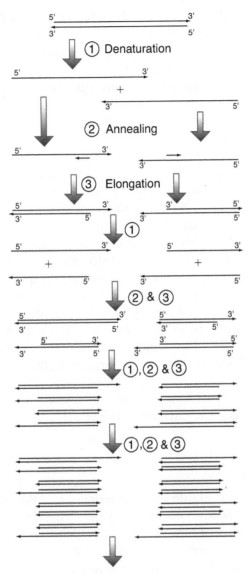

Exponential growth of short product

Figure 9-5 Diagram of PCR reaction to demonstrate how amplification leads to the exponential growth of a short product flanked by the primers. (1) Denaturing at 94–96°C; (2) Annealing at ~65°C; (3) Elongation at 72°C. Four cycles are shown here. The longer lines represent the DNA template to which primers (shorter arrows) anneal that are extended by the DNA polymerase (light circles), to give shorter DNA products (directional lines), which themselves are used as templates as PCR progresses.

antibody test could recognize maternal antibodies rather than those of the child who is exposed to HIV. It is therefore advisable to use a test that looks for the virus directly, making the same PCR test used to determine viral load in adults the best option for newborns. Infants are tested for persistent HIV antibody at about 12 months of age, but passive maternal antibody may be present until 18 months since mom's antibodies can cross the placenta and go into the baby's circulation to protect it or may come from breast milk. Detection of HIV antibody in a child after 18 months would be diagnostic of HIV infection.

In 1995, a series of papers were published on the natural history of HIV infection, which revealed that there is rapid replication and turnover of HIV, even in the clinically "latent" phase. As a result, **tests to measure viral RNA levels** were developed and became useful tools in monitoring the response of HIV to therapy in adults. These same tests also became attractive tools in the diagnosis of vertically transmitted infection. In this test, RNA is isolated from the blood sample and reverse transcription PCR (RT-PCR) is used to convert viral RNA into DNA. PCR is then used as described for the proviral DNA to amplify this DNA. This test can quantify the amount of viral RNA in the blood and can detect as little as 50 copies of viral RNA per milliliter of blood, which translates to 25 viral particles since each viral particle contains two copies of RNA. New tests continue to be developed each day and recently a Real-Time Immuno-PCR test, which combines parts of traditional antibody testing with PCR, has been developed. This new test can reportedly detect the virus when only two copies of HIV are present per milliliter of blood (www.hivandhepatitis.com/recent/test/realtime/061604f.html). This would allow clinicians to detect HIV in the blood 12 days after a person has been exposed to the virus!

Rapid HIV Tests

The major drawback of the HIV tests discussed above is that they take a long time before the results become available to the physician and the patient. Most test results become available anywhere from one to two weeks after the blood is drawn. The FDA recently approved a number of rapid HIV tests, which function on the same principle as the ELISA, but the results are available in less than one hour. Some of the major advantages of the rapid tests are that increased numbers of people benefit from knowing their HIV status; that there is an increased uptake of results by people being tested; and that less reliance is placed on laboratory services for obtaining the results.

In March 2004, the FDA approved the **OraQuick** Rapid HIV-1 Antibody Test for use with oral fluid and plasma specimens. The test was previously approved for use with blood samples in 2002. Although there are several rapid HIV tests on the market, this is the only rapid HIV test to be approved in the United States by the FDA for use with oral fluid. The OraQuick Advance test can now be used with oral fluid specimens taken from the mouth, with plasma or with whole blood. The test is performed by using the device to gently and completely swab both upper and lower outer gums one time around. The device is then inserted into a vial containing a developer solution. After 20 minutes, the test device is visually examined for the results. If HIV antibodies are present in the solution, two reddish-purple lines will be displayed in a small window in the device. Rapid tests are single-use and do not require laboratory facilities or highly trained staff. This makes rapid tests very suitable for use in resource-limited countries. It also allows testing, counseling, and referrals to be done in one visit. This test is not yet available for home use. As is true of current ELISA antibody procedures, an initial reactive rapid HIV test result should be confirmed by Western blot. The patient should be in a position to receive the necessary counseling when the results are presented and follow-up visits and care should be arranged.

Tests for HIV Progression

Various tests are available to monitor HIV disease progression and the overall health of the infected person. Tests for HIV viral load can provide a good picture of viral activity, and relative viral quantity in the circulation, while CD4 cell counts monitor the status of the immune system and may assist physicians in predicting and subsequently preventing the onset of opportunistic illnesses. A viral load test in conjunction with a CD4 cell count can help guide treatment decisions and indicate whether treatment is working.

Viral load describes the amount of HIV particles in the circulating blood and is expressed either as copies of RNA per milliliter of blood (copies/mL) or as logs. A viral load of 100,000 copies/mL or greater is considered high, while levels below 10,000 copies/mL are considered low. Research has consistently shown that higher viral loads are associated with more rapid HIV disease progression and an increased risk of death. Sometimes with successful treatment, the level of HIV may be too low to be measured, and the viral load is said to be undetectable, or below the limit of quantification. This, however, does not mean that HIV has been eradicated and the person is now disease-free. People with undetectable viral load maintain very low viral levels in circulating blood cells. Moreover, the circulating levels of HIV are much less than amounts found in immune organs such as the lymph nodes where there is a higher concentration of affected cells. In addition, even when HIV is not detectable in the blood, it may be detectable in the semen, female genital secretions, cerebrospinal fluid, and other tissues. Effective anti-HIV treatment can often reduce viral load to low or undetectable levels. Therapy that does not produce an undetectable viral load is often said to be failing, possibly because of viral mutation.

The **CD4 T-cell count** is not an HIV test, but rather a procedure where the number of CD4 T cells in one microliter of blood is counted in a standard medical lab test after a blood draw. This CD4 test does not check for the presence of HIV. It is used to monitor the immune system function in HIV-positive people by measuring the declining CD4 T cell. In HIV-positive people, AIDS is officially diagnosed when the count drops below 200 cells or when certain opportunistic infections occur. This use of a

CD4 count as an AIDS criterion started in 1992. This 200 count is not arbitrary, but was chosen because it corresponds with an increased likelihood of opportunistic infections. Lower levels of CD4 counts in people with AIDS are indicators that prophylaxis against certain types of opportunistic infections should be instituted. Generally speaking, the lower the number of T cells, the lower the immune system's function will be. This means that the body will be more vulnerable to opportunistic infections. Normal CD4 T-cell counts are between 1,000 and 1,500 CD4+ T cells per microliter and the counts may fluctuate in healthy people, depending on recent infection status, nutrition, exercise, and other factors—even the time of day. Women tend to have somewhat lower counts than men.

HIV Genotyping and Phenotyping Tests

The virus population in an HIV-1 infected individual has genetically distinct viral variants that evolve from the initial virus inoculum. For retroviruses, such as HIV-1, this is mainly due to the fact that its reverse transcriptase does not have the DNA proofreading mechanisms that seek out and correct errors as the viral genome is being replicated, coupled with rapid viral turnover. It is estimated that there are 10^9 viral particles produced per day in the body of an HIV-infected individual and about 3,300 of the newly produced viruses will carry a particular mutation each day. When antiviral drug selective pressure is applied to an infected person, preexisting minor viral species resistant to that drug rapidly become predominant and are selected for as the most fit species in the presence of the drug. Drug resistance testing of HIV-1 could be a valuable tool in patient management since resistance decreases the ability of drugs to control HIV and affects treatment options. **Genotyping** of the reverse transcriptase and protease genes has proven effective in identifying mutations associated with resistance to reverse transcriptase and protease inhibitors. The test consists of isolating the viral particles from a patient and using advanced DNA technology to sequence the HIV-1 protease and reverse transcriptase genes and to predict the potential drug resistance based on the deduced amino acid sequences of these proteins. **Phenotyping** examines the ability of a patient's HIV-1 to replicate in vitro in the presence of different concentrations of antiretroviral drugs. Phenotyping helps predict viral load response to new antiretroviral drugs. Performing drug resistance tests up front may provide the most valuable information to maximize the effect of any combination antiretroviral treatment. Resistance tests are helpful when choosing a drug regimen. The tests are only a guide, however, other factors, such as past medications, side effects, and adherence must be taken into account as well. There are test kits available for these tests. Examples include *PhenoSense*^{TM} *HIV* made by ViroLogic Inc., *VircoGEN*^{TM} made by Virco/ Laboratory Corporation of America, and *TruGene*^{TM} from Visible Genetics. These tests are not routinely performed at hospital labs and would most likely have to be sent to a reference lab with 10-day to two-week turnaround.

Test for HIV Tropism

Tropism describes the mechanism used by HIV to infect CD4+ cells. As discussed in Chapter 4 under the sub-heading "Cellular Receptors and Entry of HIV," in order to enter a CD4+ T cell, HIV first attaches to the CD4 receptors on the surface of the cell. However, the virus cannot initiate membrane fusion and entry until it gets closer to the cell membrane. In addition to the CD4 receptor, HIV needs to bind to either the CCR5 or the CXCR4 co-receptor susceptible on cell surfaces in order to gain entry. Determining the tropism of HIV is important because there is a class of HIV drugs that work against CCR5-tropic HIV, but have no effect on CXCR4 ones. These drugs are called CCR5 antagonists or entry inhibitors and include drugs such as Pfizer's Selzentry (generic name, maraviroc). These antagonists work by attaching to the CCR5 co-receptor on the CD4 cell surface, thus blocking HIV fusion and entry.

Routine HIV Testing as Part of Regular Medical Care

The CDC estimates that between 20 and 25% of the roughly 1.3 million HIV-infected Americans are unaware of their status and therefore are unknowingly transmitting the virus. Current studies indicate that HIV-infected persons access the healthcare system but are not tested for HIV until late in the course of their disease. For many, it is too late to benefit from the effective HIV treatments that are available today. According to the latest CDC update on HIV testing in the United States, 32% of persons with HIV infection diagnosed in 2009 progressed to AIDS within a year, which indicates those infected may have been living with HIV for up to 10 years before being diagnosed and having access to HIV medical care. In 2006, the CDC therefore recommend that all Americans between ages 13 and 64 be tested for HIV at least once during their lifetime. The CDC recommends that testing should be made a standard part of medical care, regardless of individual risk factors or HIV prevalence in the community. These recommendations are a result of extensive consultation with representatives from professional organizations, state and federal agencies, community-based organizations, people living with HIV/AIDS, health departments, and academia.

Knowledge of one's status allows for earlier access to effective treatment that results in a longer, healthier life span. When people learn that they are HIV infected, the necessary steps can then be taken to protect their partners. In the past, knowledge of positive HIV status has been shown to reduce risky behavior with uninfected partners by approximately 68%. The CDC estimates that most of the new HIV infections in the United States are transmitted by the 20% of people who do not even realize that they are infected. The hope is that by expanding access to HIV testing and making it a standard part of medical care, those who are not aware that they are infected may be effectively reached and will in turn take steps to protect their partners. Some argue, however, that while broader testing is necessary, it is not enough to stem the tide of the HIV epidemic in our population.

Name: _____

1. True or False: There are currently several FDA-approved home testing HIV kits that can be administered by patients in the privacy of their homes and results are available in less than 10 minutes.

2. True or False: The Western blot is the standard confirmatory test for HIV.

3. _____ is the agency of the United States government that licenses HIV diagnostic tests.

4. Which of the following is the most commonly used laboratory test to detect HIV?
   (A) ELISPOT
   (B) ELISA
   (C) PCR
   (D) Northern blot

5. If you suspect that you had a high-risk encounter and could have been exposed to HIV, on average, how long should you wait before going to the doctor to get a routine HIV test?
   (A) Five days
   (B) One week is optimum
   (C) Four weeks
   (D) Six months
   (E) You should get the test as soon as possible

6. How quickly can you get the results from an HIV rapid test?
   (A) Within 30 seconds
   (B) Within 30 minutes
   (C) After 1 hour
   (D) By the end of the business day
   (E) The next morning

7. Fourth-generation HIV tests such as the Abbott Architect measure which of the following in the patient serum?

   (A) Viral RNA

   (B) HIV antigen and antibodies

   (C) HIV antigen only

   (D) HIV antibodies only

   (E) HIV-infected cells T-cells

8. Why is the viral load test better for detecting HIV in newborns than the antibody tests?

9. What is the benefit of performing an HIV genotype and phenotype test?

10. What is the role of CCR5 in HIV infection and how do CCR5 antagonists work?

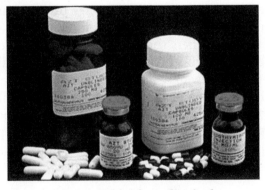

CHAPTER 10

HIV/AIDS Treatment

By 1986, AIDS researchers could describe the nature and shape of HIV proteins as well as key molecules on the immune system cells that HIV infects. Researchers began to test drugs already approved to treat other conditions to see if any of them would have an effect on the virus. They soon discovered Azidothymidine (Zidovudine, see Figure 10-1), known as AZT, that was capable of abrogating the action of the reverse transcriptase enzyme used by HIV to change its RNA into DNA so that the provirus DNA may integrate into the host genome. Currently, there are 31 FDA-approved antiretroviral drugs available to treat HIV and many more in the pipeline. In 2011, for the first time, a majority (54%) of people eligible for antiretroviral therapy in low- and middle-income countries were receiving it. There is, however, much more to be accomplished in the arena of HIV/AIDS therapy before we can start to celebrate.

The blocking of reverse transcriptase significantly slowed the progression of the virus. Antiretroviral (ARV) therapy is the main type of treatment for HIV or AIDS. The goal of anti-HIV treatment is to reduce the amount of HIV in the body. It is not a cure; however, it can significantly slow the progression to AIDS and millions of people take these drugs each day. At the end of 2012, the number of people accessing antiretroviral therapy

Figure 10-1 AZT (**zidovudine**), the first medication shown to be effective against HIV.

113

was approximately 8 million of the 14.8 million eligible patients. Eighty percent of HIV-positive individuals living in 10 of the most affected low-middle income countries were receiving antiretroviral medications, a 63% increase over the past three years. However, 7 million adults who currently need therapy and 72% of children living with HIV who are eligible for treatment still do not have access.

Nucleoside and Non-Nucleoside Inhibitors

There are two types of reverse transcriptase inhibitors—nucleoside and non-nucleoside—which target slightly different parts of the enzyme. **Nucleoside reverse transcriptase inhibitors (NRTIs),** such as zidovudine (AZT), are similar in structure to nucleosides, which are the building blocks of nucleic acid. These drugs differ slightly from nucleosides (analogs); hence, if they become incorporated into the growing DNA strand, they will interfere with the DNA replication process. Specifically, NRTIs incorporated into the growing DNA strand terminate further strand elongation. The enzyme responsible for human cellular DNA replication, DNA polymerase, is generally able to differentiate between the analogs and the real nucleotides, therefore avoiding incorporation into our DNA. The HIV reverse transcriptase is much less picky and routinely utilizes these analogs when converting viral RNA into DNA. Viral replication is therefore blocked.

Non-nucleoside reverse transcriptase inhibitors (NNRTIs), such as tenofovir DF (Viread) and stavudine (Zerit), are a chemically diverse class of drugs that bind a pocket near the active site of reverse transcriptase and as such inhibit the enzyme. The viral genetic material therefore cannot be incorporated into the genetic material of the cell, resulting in a lack of virus production. Since we do not normally have reverse transcriptase in our system, these drugs do not disrupt cellular functions. The action of NRTIs and NNRTIs on HIV was short lived. The virus soon came back to life as if the drug was not even present. The virus had mutated and the drug no longer had an effect. Even when these reverse transcriptase inhibitors were used together, they had no effect on the mutated form of the virus and the health of patient soon deteriorated.

Researchers soon came up with the idea that a multipronged approach might be best since this had worked for other rapidly mutating microbes like tuberculosis. By 1996, a new drug target was found. They devised a way to inhibit the action of another HIV enzyme called protease. Protease inhibitors stop the virus from forming mature virions, or individual virus particles by blocking the action of the enzyme on the new proteins that are formed, making the virus particles nonfunctional. Protease inhibitors [lopinavir/ritonavir (Kaletra)] seem to work miracles when used in combination with two reverse transcriptase inhibitors.

Entry Inhibitors

All reverse transcriptase inhibitors and protease inhibitors act on HIV once it enters the cell. Everyone would agree that it would be better to prevent HIV entry in the first place and, consequently, there has been a concerted effort to develop entry inhibitors.

Entry inhibitors work by attaching themselves to proteins on the surface of T cells or proteins on the surface of HIV, thus preventing entry into healthy cells of the body. In order for HIV particles to bind to T cells, the proteins protruding through the viral envelope must bind to the proteins on the surface of the T cells. Entry inhibitors would prevent this from happening. Some entry inhibitors target the gp120 or gp41 proteins on the surface HIV, while others target the CD4 protein, the CCR5 or CXCR4 receptors on the surface of T cell, and other susceptible cells in the body. If entry inhibitors are successful in blocking these proteins, HIV is unable to bind to the cell surface and gain entry. Fusion inhibitors, such as enfuvirtide (Fuzeon or T-20), work by blocking HIV entry into cells. Fuzeon is a peptide that binds to gp41, preventing HIV from binding to the surface of T cells. It was approved by the FDA in 2003. There are other experimental drugs that show early promise as well. PRO-542 and TNX-355 target the CD4 protein, and vicriviroc and maraviroc target the CCR5 protein used as coreceptors to enter cells.

Integrase Inhibitors

A critical step in the HIV life cycle is the integration of the virus's genetic information into the host cell DNA. This process enlists the host cell as a "HIV factory," which is programmed to produce numerous virions each hour. Integrase is the enzyme that accomplishes this task. Like proteases and reverse transcriptase, integrase also plays a vital role in the retroviral life cycle, which makes this enzyme an attractive target for the development of new anti-AIDS agents. There have been several attempts to create integrase inhibitors over the past 10 years. Successful integrase inhibitors are oligo-nucleotides, which are small segments of DNA or RNA that are synthetically prepared. Modified oligonucleotides can serve to block RNA/DNA interactions, thus modifying protein or enzyme synthesis. One drawback to integrase inhibitors is that they only have one chance to act for each cell. If an inhibitor fails, any further attempts are futile since the genetic information would already be incorporated. Raltegravir (Isentress™) is a drug from Merck & Co. that targets integrase. It was the first integrase inhibitor to go to trial and received FDA approval in 2007. In December 2011, it received FDA approval for pediatric use in patients ages 2–18. In March 2013 data reported at the the 20th Conference on Retroviruses and Opportunistic Infections (CROI2013) demonstrated that a new integrase inhibitor (dolutegravir) was more beneficial than raltegravir for treatment of experienced people with resistance to two or more anti-retroviral drug classes. Dolutegravir is taken once daily, was generally safe and well tolerated in trial participants, and had lower rates of virological failure. In February 2013 the FDA announced that it would fast track dolutegravir's approval process, to the delight its developers, ViiV/Shionogi in partnership with GlaxoSmithKline. Being able to add another effective drug to the current combination therapy is certainly good news for HIV-positive patients.

HAART

With the new knowledge that the virus can mutate and that multiple drugs might need to be administered at the same time to get the desired effect, **highly active antiretroviral therapy** (HAART) became the recommended treatment for HIV infection in 1996. This is also called **combination therapy** (Figure 10-2 shows selected antiretrovials). HAART combines three or more anti-HIV medications in a daily regimen, sometimes referred to as a "cocktail," thereby reducing the chances of viral mutation against all of these drugs.

So far, the combination HAART treatment is the closest thing medical science has to an effective therapy. The key to its success in some patients lies in its ability to disrupt HIV at different

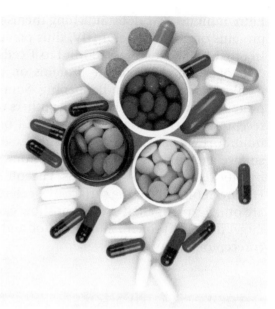

Figure 10-2 A selection of HIV/AIDS antiretroviral drugs.

stages in its replication cycle. Reverse transcriptase inhibitors, which usually make up two drugs in the HAART regimen, inhibit an enzyme crucial to the early stage of HIV replication (RNA to DNA). Protease inhibitors hold back another enzyme that functions near the end of the HIV replication process. HAART has been credited as a major factor in significantly reducing the number of deaths from AIDS, especially in the United States and other developed countries where access to treatment is not restricted by cost. The World Health Organization recommendations for HIV treatment state that three separate antiretroviral medicines need to be taken at all times. While HAART is not a cure for AIDS, it has greatly improved the health of many people with HIV, and it reduces the amount of virus circulating in the blood to nearly undetectable levels. Researchers, however, have shown that HIV remains present in hiding places, such as the lymph nodes, brain, testes, and retina of the eye, even in people who have been treated.

Although HAART has several beneficial effects, there are several side effects associated with the use of these drugs, and many of these effects may be severe. Some of the NRTIs may cause a decrease of red and/or white blood cells (pancytopenia), especially when taken in the later stages of the disease. Some may also cause inflammation of the pancreas and painful nerve damage. This reduction in red blood cells can lead to anemia, increased bruising or bleeding, and an inability to efficiently clot the blood. These latter effects stem from the fact that the blood platelets that are responsible for the initial stages of the clotting process are greatly reduced. There have been reports of complications and other severe reactions, including death, to some of the antiretroviral nucleoside analogs when used alone or in combination. Therefore, any patient taking these drugs should be closely monitored for signs of these side effects. Protease inhibitors also have associated side effects. The most common side effects associated with protease inhibitors include diarrhea and nausea. In addition, there is also a risk of drug interactions when taking protease inhibitors, which may result in serious side effects. Fuzeon inhibitors may also cause severe allergic reactions such as pneumonia, labored breathing, chills, fever, skin rash, blood in urine, vomiting, and low blood pressure. There have also been reports of local skin reactions at the injection site (it is given as an injection underneath the skin). Researchers and physicians alike have however determined that the benefits outweigh the risks associated with treatment using HAART and other anti-HIV drugs. AIDS killed an estimated 1.7 million people in 2011. If everyone had access to ARV therapy, then the death toll would be much lower.

Atripla and AIDS Treatment

On July 12, 2006, the FDA approved Atripla tablets, a fixed-dose combination of three widely used antiretroviral drugs in a single tablet taken once a day, alone or in combination with other antiretroviral products for the treatment of HIV-1 infection in adults. Atripla, the first one-pill, once-a-day product to treat HIV/AIDS, combines the active ingredients of Sustiva (efavirenz), Emtriva (emtricitabine), and Viread (tenofovir disoproxil fumarate). Each component of Atripla is currently approved for use in combination with other antiretroviral agents to treat HIV-1-infected adults. The safety and effectiveness of each component were previously demonstrated in clinical trials to support their individual approval. In addition, the safety and effectiveness of the combination of these three drugs were shown in a 48-week-long clinical study with 244 HIV-1-infected adults receiving the drugs contained in Atripla. The approval of Atripla simplifies the treatment regimen for HIV-1-infected adults, and will potentially improve the ability of patients to adhere to treatment resulting in long-term effective control of HIV-1. In the clinical trial phase, 80% of the participants achieved a marked reduction in HIV viral load and a substantial increase in the number of healthy CD4 cells needed to combat the infection. It is hoped that Atripla will significantly simplify the drug treatment regimen, helping to reduce pill burden, increase adherence, thus reducing potential development of viral resistance to the drugs. Like the individual drugs that make up this single dose therapy, Atripla is not without side effects. The most common side effects include headache, dizziness, abdominal pain, nausea, vomiting, and development of a rash. Other potentially serious adverse events reported for the use of Atripla's ingredients include liver toxicity, renal impairment, and severe depression in a subset of patients (www. fda.gov/cder/drug/infopage/atripla/factsheet.htm). The drug currently retails in the United States for US$1,850 for a one-month supply.

Functional Cure for HIV

There are mounting opinions suggesting that it is possible to control the virus that causes AIDS with early treatment, so further therapy is not immediately needed. A functional cure can be defined as "long-term health in the absence of therapy." There might be some virus left in the body, but it's not causing any damage. The first functionally cured HIV patient, Timothy Ray Brown, dubbed the "Berlin Patient," was HIV positive when he also developed a serious case of acute myeloid leukemia, unrelated to the HIV infection. Brown's oncologist, Dr. Gero Hutter of the Charite Hospital in Berlin, decided that his patient needed a bone marrow transplant. Brown's doctor decided to find a bone marrow donor who carried the CCR5-Delta 32 homozygous mutation, which is present in approximately 10% of people of European descent. They were lucky to find a match, and Brown received two bone marrow transplants in Berlin, in 2007 and 2008 following high doses of chemotherapy and radiation. The transplant not only effectively cured Brown's leukemia but also eliminated the HIV from his system. He stopped taking his antiretroviral medication the day of the transplant, and the last direct evidence of HIV was detectable on day 60 after his transplantation. Brown has been off antiretroviral therapy for about five years now, and there is no evidence of the virus rebounding. In fact, his antibody levels have been declining steadily, suggesting that the virus is not present in the numbers required to continue stimulating the immune system.

In a recent article published in the journal *PLOS Pathogens*, researchers, from the Institut Pasteur in Paris showed that early and prolonged combination antiretroviral therapy (cART) for their HIV infection may allow some individuals to achieve long-term infection control. In fact, the group reported that to date, 14 adults have been functionally cured of HIV after been given cART. The report confirms that all patients have been able to stop taking antiretroviral treatment while still keeping their infection under control. Overall the data suggests that approximately 5 to 15% of individuals who started receiving cART between five and 10 weeks of becoming infected are able to control the virus after stopping all antiretroviral treatment, essentially becoming functionally cured. On average, the patients in the study were off their medications for over seven years at the time of publication. The fact is that no one really knows if the virus will rebound in some of these individuals at some point, but the experiences of these patients renew hope that a cure for HIV is possible someday.

This hope for a cure was bolstered in the spring of 2013 when a baby born at 35 weeks to a Mississippi mother who didn't realize she had HIV until she came into the hospital during labor, was aggressively treated with combination HIV drugs within 30 hours of birth. The doctor, who treated this baby girl—Dr. Hannah Gay, associate professor of pediatrics at the University of Mississippi—felt that the baby was at higher than normal risk because of the lack of prenatal care of the mother and started treatment even before the HIV test results for the baby were obtained. The repeated test results confirmed that the baby was HIV positive, and she therefore continued to receive aggressive antiretroviral treatment. The viral load rapidly declined with treatment and fell from around 20,000 copies/ml of blood to undetectable by the time the baby was a month old. The levels remained undetectable until the baby was 18 months old, after which the mother stopped coming to the hospital for continued treatment. Five months later when Dr. Gay finally located the mother, she expected to see high viral loads when the baby was retested. However, to her surprise, the tests were negative! Retesting was ordered, and they all came back negative. Further testing with a more sensitive method performed by Dr. Katherine Luzuriaga, an immunologist at the University of Massachusetts Medical School in Worcester, MA, found minute amounts of some viral genetic material but no virus able to replicate, even in the typical viral reservoirs in the body. Now at almost age 3, the child has been off drugs for over a year and a half with no signs of replicating HIV particles.

In the case of this Mississippi baby, the circumstances leading to a functional cure were certainly unconventional. The most likely scenario leading to a functional cure in this baby is that the aggressive early drug therapy killed off the HIV before it could establish a hidden reservoir in the baby's lymph nodes, spleen, and bone marrow. The major reason most individuals have to keep taking anti-HIV drugs today is that the virus hides in a dormant state, out of reach of existing drugs. When drug therapy is stopped, the virus can emerge from hiding and again seed the circulating blood, rapidly increasing the viral load and causing immediate damage to infected CD4 T cells. Therefore, drugs designed to cause the virus to leave these hiding places in conjunction with cART would go a long way in helping others to achieve a functional cure.

Bee Venom and HIV Treatment

In March 2013, scientists from Washington University School of Medicine reported that they had encapsulated bee venom toxin melittin in nanoparticles and demonstrated its ability to destroy HIV and spare uninfected cells. Melittin is a potent toxin found in bee venom and previously shown to have the ability to poke holes in the viral envelope that surrounds many viruses, including HIV. Melittin attacks double-layered membranes indiscriminately, significantly broadening its range of activity. In fact, free melittin can cause considerable damage to not only viral particles but also mammalian cells. Senior author of the report, Samuel A. Wickline, had previously demonstrated that nanoparticles loaded with melittin have anti-cancer properties, effectively killing tumor cells. This group of scientists, including Joshua L. Hood, obtained funding from the Bill & Melinda Gates Foundation Grand Challenges Explorations. They used the funding to construct a set of nanoparticles and coated their surfaces with melittin. They then placed "molecular bumpers" on the surface of these nanoparticles that allowed them to bounce off normal cells when they encounter them. HIV, which is significantly smaller than a human cell, is able to fit between the bumpers and make contact with the melittin on the surface of the nanoparticles, leading to the destruction of the viral membranes, literally stripping it off the virus (Figure 10-3). This type of attack on HIV is very different from current anti-HIV drugs, which all attempt to inhibit the ability of the virus to replicate. As we well know, HIV evolves to evade this type of treatment. However, melittin attacks the inherent structure of the virus and has the ability to kill HIV at the inception of the infection, even prior to entry into a host cell. This is an exciting breakthrough since several other viruses, such as hepatitis B and C, have similar types of envelopes and would also be susceptible to the toxic effects of melittin. This formulation is being considered for incorporation into a vaginal gel to prevent the spread of HIV infection.

Treatment Outlook

We do not yet know how long people can live with HIV infection if they are tested early and treated appropriately. We currently have very potent combination antiretroviral medications against HIV, but these have only been in use since the late 1990s. That said, the current outlook for effective HIV treatment has never looked better, especially for those who begin anti-retroviral therapy at an early stage of the disease. The AIDS-related death rate in some parts of the developing world, however, remains staggeringly high due to lack of access to these life-saving drugs. A recent study conducted in South Africa confirms that the use of potent combination antiretroviral therapy can dramatically reduce the morbidity and mortality associated with HIV infection even in developing parts of the world. In the developed world, the age-adjusted average life expectancy of patients living with HIV is now estimated to be close to that of their

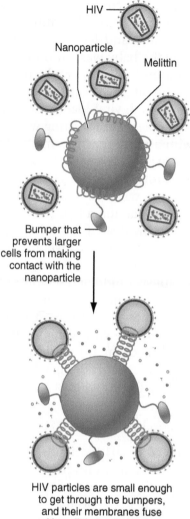

HIV particles are small enough to get through the bumpers, and their membranes fuse with melittin leading to lysis of the viral particles.

Figure 10-3 Model of bee venom nanoparticles used to destroy HIV through a direct membrane attack.

uninfected peers. Using survival modeling, the researchers calculated excess mortality attributable to HIV after treating patient cohorts with ART. Life expectancy for HIV-infected patients on ART was approximately 80% of the normal expectancy. The results varied by age at ART initiation, baseline CD4-cell count, and sex. Essentially, the younger the age at initiation, higher baseline CD4-cell count, and female sex were associated with the greatest benefit.

New drug therapies and new trials are being developed every day in the labs of a diverse researchers, some of whom have dedicated their entire careers to HIV research. Those who have experienced the rapid developments in this field over the last three decades will admit that it has been quite a roller-coaster ride. The dream of a cure or at least eradication of the infectious virus from the body of an infected person that was widely hoped for in the beginning eventually faded as trial after trial failed. In 1997, it was estimated that viral suppression using antiviral drugs for a duration of three years was necessary to completely eradicate the virus from an individual. Researchers arrived at this conclusion by a combination of mathematical models and lab observations. The thought was that after this period, all infected cells would presumably have died. Since then, the duration of three years has consistently been adjusted upward and newer studies came to the sobering conclusion that HIV remains detectable in latent infected cells, even after long-term suppression.

To date, nobody knows how long these latent infected cells survive, and all indications are that even a small number of them would be sufficient for the infection to flare up again as soon as treatment is interrupted. The most recent estimates for eradication of HIV-infected cells from the body are approximately 50–70 years. However, there has been significant knowledge gained over the past five years that has brought renewed hope to HIV treatment. There remains, however, significant challenges that must be overcome before consistent and reliable ablative treatments will be realized. Dr. Michael Gottlieb described it best in a *Los Angeles Times* article when he said, "I've always looked at AIDS therapy as a series of leaky lifeboats. You stay in the first one until you're sinking, then jump to another one. But you don't give up looking for others." HIV/AIDS has not been a respecter of anyone. It has crossed racial, gender, socioeconomic, religious, regional, and continental lines. While it is true that the disease has been devastating in areas such as sub-Saharan Africa, it is also very prevalent in MSM and injection-drug users in the United States, and literally in all of our backyards. Instead of eradication, it has become more realistic to consider the lifelong management of HIV infection as a chronic disease in a similar manner to diabetes mellitus. This will, however, mean that patients will need to have the discipline and ability, both mentally and physically, to take the currently available therapy, sometimes more than once daily at fixed times for the next 10, 20, or even 30 years. Regardless, HIV remains a dangerous and cunning opponent on which we cannot afford to turn our backs. The developments in the area of treatment and management of the disease over the next five to ten years will be very important in shaping the future of this deadly epidemic.

Test Your Knowledge

Name: _____

1. Which of the following classes of HIV drugs prevent HIV replication by binding to the reverse transcriptase enzyme and preventing its action?
   (A) NRTIs
   (B) PI
   (C) FI
   (D) NNRTIs

2. People infected with HIV must take drug combinations, which means taking several tablets each day. In 2006, the FDA approved a once-a-day pill called _____ that combines three drugs in a "cocktail" therapy that can be swallowed in a single dose once a day.
   (A) AZT
   (B) Atripla
   (C) maravirorc
   (D) raltegravir

3. The antiviral drug called Fuzeon works by _____.
   (A) binding to Gp41 and preventing HIV entry
   (B) preventing Gp120 from binding to CD4
   (C) binding to CCR5 on T cells, preventing HIV entry
   (D) blocking HIV envelope development

4. What does an HIV viral load test measure?
   (A) The average weight of a viral particle
   (B) The amount of virus in the blood
   (C) The amount of virus in the bones
   (D) How long you have been infected

5. What was the unique characteristic of the bone marrow transplanted into the Berlin Patient, Timothy Brown, which resulted in functionally curing his HIV?

    (A) The donor had a homozygous CCR5 mutation.

    (B) The donor's T cells were devoid of CD4 receptors.

    (C) The donor's blood produced antibodies that killed the HIV.

    (D) The donor's macrophages did not recognize HIV p24 protein.

6. True or False: Approximately 5 to 15% of individuals who started receiving cART early after infection are able to control the virus after stopping all antiretroviral treatment, essentially becoming functionally cured.

7. True or False: The majority of people eligible for antiretroviral therapy in low- and middle-income countries are currently receiving it.

8. True or False: The World Health Organization recommendations for HIV treatment state that at least two separate antiretroviral medicines need to be taken at all times

9. What is the most likely explanation for the apparent functional cure of the Mississippi baby?

10. What is the name of the bee venom toxin that was recently demonstrated to destroy HIV particles and what is its mechanism of action?

CHAPTER 11

HIV Vaccine Research: Is There Real Progress?

Current Perspective

Throughout human history, no major viral epidemic has ever been defeated without an effective vaccine. Reflecting to 1984 in the midst of the exciting announcement that the cause of AIDS had been discovered, federal health officials famously predicted that they would have a vaccine ready for market within three years. Back then it was rightly recognized that AIDS was indeed a deadly disease with the potential to wreak havoc across the planet. These very experienced physicians and scientists also knew that historically, vaccines have been the most effective weapon against the world's deadliest infectious diseases. No one could have anticipated how formidable a foe HIV would prove to be. It was soon discovered that HIV has an array of unique ways that it can evade the human immune response and that the human body seems incapable of mounting an effective immune response that would eliminate the virus, even with help from the best combination of antiretroviral drugs. Consequently, scientists still do not have a clear understanding of what is needed to provide protection against the virus. Now, over three decades later, the landscape of HIV vaccine research is littered with signs of valiant efforts, tremendous struggles, epic disappointments, and intermittent waves of optimism.

Even with the great success and increasing availability of antiretroviral drugs, which have been shown to significantly extend and improve the quality of life of HIV-infected persons, these drugs do not "cure" the HIV-infected patient. Physicians and their patients alike would be much happier about an intervention that is capable of completely preventing HIV infections. While condom use, abstinence, and needle-exchange programs have been very effective in significantly reducing the rate of HIV infection in many areas over the past decade, these practices are totally dependent on responsible cooperation of all sexually active individuals. As we all know, human behavior is not easily controlled, and people will do whatever they want despite the possible consequences. A strategy that would completely prevent HIV infection is therefore highly desirable.

Almost every time I talk with students, family members, and friends about HIV infection, viral mutation, drug resistance, the prevalence of the disease, and the implications for the next generation, I get asked the all-important question, "How far are we away from making an HIV vaccine?" The fact is that there is currently no vaccine to prevent HIV. Researchers, however, continue to develop and test potential HIV vaccine targets with the hope that they will in the near future develop a vaccine that can protect people from HIV infection, or at least lessen the chance of getting HIV or develop AIDS should a person be exposed to the virus.

How Do Vaccines Work?

What is a vaccine anyway and why is it such a desirable treatment option for HIV/AIDS? A vaccine is a medical product that is designed to stimulate the body's immune system in order to prevent or control an infection. An effective preventive vaccine primes the immune system to fight off a particular microorganism so that it can't establish a serious infection or make you sick. A vaccine preparation therefore contains an antigen that may consist of whole disease-causing organisms (killed or weakened) or parts of such organisms, which is used to confer immunity against the disease that the organisms cause. Vaccine preparations can be natural, synthetic, or derived by recombinant DNA technology. These vaccines are typically delivered orally, intramuscularly (see Figure 11-1.), nasally or transcutaneously.

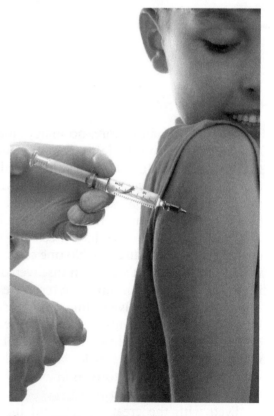

Figure 11-1 An effective preventative vaccine against HIV is the great hope for this century.

There are three main ways to make a vaccine, although most strategies for an HIV/AIDS vaccine focus on a single strategy.

One can make what is known as a **killed virus vaccine**. This would be prepared by taking HIV particles and "killing" them with chemical, UV light, or heat treatments. This killed product is then mixed with an agent that helps to boost the immune system. While this strategy has been successful in the past in creating vaccines like the Jonas Salk polio vaccine, so far it has not been successful in AIDS vaccine development. A second strategy would be to make a **live attenuated vaccine**. In this case the viruses would still be alive, but would be weakened in order to reduce their ability to cause disease, but just enough to still allow it to induce immune responses inside the vaccine recipient. As you might imagine, with this approach there is the danger of getting HIV infection from the vaccine and, consequently, this approach has not been explored. As Dr. Emilio Emini, Senior Vice President and Chief Scientific Officer of Vaccine Research at Pfizer Inc (former head of vaccine development at Wyeth), explains, "In the case of live attenuated HIV, that virus never goes away because it becomes an integral part of the genetic information of the cells in that individual. Eventually it will cause disease." This does not mean that researchers have run out of options quite yet.

The third approach to vaccine development is the most recent and the safest of the three. It employs a **genetically engineered** or **subunit vaccine**. Subunit vaccines do not involve the use of the whole virus particles, but are constructed with two or more viral proteins. With this approach, vaccinologists can manipulate the nucleic acid (genetic material) of the pathogen in several ways. Certain integral genes whose products stimulate an immune response may be inserted into some sort of vehicle, typically another virus that does not cause disease in humans. This genetically engineered virus is injected into the person where it triggers an immune system. This live vector vaccine is able to stimulate a much stronger immune response than killed viruses. Alternately, they could create a vaccine by injecting only certain elements of the virus, such as portions of its glycoproteins, or synthesized pieces of viral protein. The bottom line in this approach is that the immune system would recognize and respond to these viral components, building up memory B and T cells ready to respond in the event that a real HIV particle is introduced to the body. The downside to this approach is that the immune system does not get to sample the entire virus, but is limited to the components in the vaccine mixture. This approach is, however, very safe. Most current HIV/AIDS vaccine strategies employ genetically engineered vaccine strategies.

Real Progress or Fool's Gold?

Despite the enormous challenges highlighted, there have been cause for great optimism over the past five years. Researchers identified, isolated, and closely analyzed dozens of exceptionally potent antibodies that demonstrated the ability to neutralize a vast number of HIV variants from around the world. They have now carefully examined the mechanisms underlying how these antibodies bind to HIV and block its entry into target cells. Many believe that these new findings establish a foundation for the development of potentially powerful AIDS vaccines with preventative capabilities.

This optimism was bolstered in 2009 when a phase III clinical trial of an HIV experimental vaccine demonstrated for the first time that a vaccine could prevent HIV infection. The study was conducted in Thailand and consisted of two previously unsuccessful vaccine constructs, ALVAC (Gag, Pol, and gp160), from Sanofi Pasteur and AIDSVAX (gp120), originally developed by VaxGen Inc. The study enrolled a total of 16,402 Thai male and female volunteers between age 18 and 30 years old who were HIV negative. The volunteers were divided into treatment and a placebo group. Study participants were given six shots of the experimental vaccine over six months. Initially, they received two of Alvac followed by two more sessions where they were injected with both vaccine preparations. They were then tested for HIV every six months. At the end of the trial only 125 study volunteers had become infected with HIV. New infections occurred in 51 of the 8,197 given vaccine compared to 74 of the 8,198 who received dummy shots. That worked out to a 31% lower risk of infection for the vaccine group. The data apparently came as a surprise to the joint U.S. and Thai researchers, making headlines on almost every newscast across the globe. It should however, be noted here that while it was not emphasized in the hyped media reporting, the vaccine did not protect those at high risk of HIV infection, such as sex workers and intravenous drug users, and the protective effect was greatest in the first 12 months and then seemed to diminish over the remainder of the study period. Finally, when those who did not get all six vaccine shots were taken out of the analysis, the positive result was statistically insignificant. I will hasten to state, however, that although the protection provided by this vaccine was too modest to support licensure, subsequent analysis of the immune responses induced by the vaccine regimen has provided valuable data and new insights that will no doubt be applied to the design and clinical evaluation of future HIV vaccine candidates.

In the process of designing the numerous clinical trials for various HIV vaccine formulations over the past decade, researchers now have novel and exciting tools that have significantly increased our capabilities to better understand the infection mechanisms, revealing replication vulnerabilities that can now be exploited. Moreover, new and more effective delivery systems for these vaccine antigens have been developed and are currently being tested for their efficacy to safely deliver HIV vaccines to humans. This has reinvigorated the waning optimism in the field about the prospects for the development of a safe and effective AIDS vaccine, despite the longstanding challenges.

The Challenges of Creating an HIV Vaccine

It is no secret that vaccine development has always been a long and arduous process. However, several factors make HIV a particularly challenging virus to create a vaccine against. The main problem that HIV vaccine developers face is the unusual nature of HIV. The virus not only evades the body's immune system response, it actually destroys the very immune system cells (the CD4 T cells) that are central to the body's defense against a viral or bacterial infection. As soon as the virus gains entry into the CD4 cells, it inserts its genetic material into the cell's genome and literally hijacks the cells protein-making machinery to create more viruses. If a vaccine is able to stimulate the immune response and the virus is not being killed, this could actually spur further virus production. To compound an already bad situation, HIV, like all other retroviruses, must convert its genetic code from RNA to DNA once inside the CD4 cell. It turns out that this process is notoriously error-prone. These translational errors often lead to high viral mutation rates. Since the immune system typically defends the body by producing highly specific antibodies (see Figure 4-2), mutated viruses would not be recognized and would therefore be able to escape the immune response stimulated by a genetically engineered subunit type vaccine. The vaccine-resistant virus strains could therefore be transmitted to others. Within HIV-1 alone, there are at least 10 different subtypes (as well as recombinant viruses combining several subtypes), which are found in different parts of the globe (see Figure 3-3. Researchers must therefore develop multiple vaccines or one that works optimally across multiple subtypes.

Even if researchers could solve all of the problems and work out all of the issues cited above, they would still have to face the greatest challenge yet in creating an efficacious AIDS vaccine; *that is, in the past 33 years since the AIDS pandemic started, doctors have not found a single patient who was able to rid himself/herself of the virus.* This is an extremely discouraging problem facing vaccine researchers today. To date, even with the deadliest of diseases caused by viruses such as smallpox, polio, measles, or bacterial infections caused by *Salmonella* species, *Mycobacterium tuberculosis*, and *Escherichia coli* (with only a few exceptions—herpes), when one gets infected, there's a certain level of pathology as well as mortality in some individuals; then you clear the infection, the body mounts an immune response, and it's unlikely you're going to get infected again. Once you get infected with HIV, the body seems to be completely incapable of eliminating all the virus particles. Over the years, successful vaccine development has always taken its cue from what the body did to protect itself under natural circumstances. Researchers then develop a vaccine that mimics what the body has done in order to provide some initial help to the system so it does not get over-whelmed. With HIV the body has not been successful in mounting a response that clears the virus from the system; therefore, there is no real framework on which to base vaccine development. This does not mean that the immune response is nonexistent; far from it. The body mounts a robust response and is in part responsible for the pre-dictable latent period seen in HIV infection where the viral load in many patients is almost undetectable, even without treatment. The problem is that this type of control is not sustainable and HIV eventually gains the upper hand, having destroyed a key immune system player, the T-helper cells. Despite these nearly insurmountable chal-lenges, the discovery and development of safe, efficacious, and cost-effective vaccines to prevent HIV infection and/or disease has become the mission of several research groups worldwide (www.iavi.org).

Preventive vs. Therapeutic Vaccines

When most people think about a vaccine against HIV, what they are envisioning is a product that will prevent infection in the first place since the virus cannot be cleared. The fact is that there is no vaccine available for any disease that prevents the organism from entering the body or cells in the first place. Vaccines typically prevent disease and not infection.

A **preventive vaccine** is designed to block infection in people who are HIV negative. When our bodies encounter a microbe of any type, the immune system responds with all its arsenals with two important goals—first, to eliminate the threat, and second, to build an immunologic memory so that it continues to "remember" how the microbe looked and can quickly defeat the invader should it ever try to infect again. Since the vaccine is designed to look like the real microorganism, it trains the immune system to recognize and attack the real microorganism should it ever be encountered by the vaccinated individual. Therefore, if you have received an effective protective vaccine, your immune system will "remember" how to attack and quickly defeat that particular microorganism for many years. The principle here for preventive vaccine success is the induction of antigen-specific B and T cells that, in addition to having effector clones, also produce memory cells that are long lived and are able to be rapidly reactivated if that antigen is seen again (Figure 11-2). The majority of early HIV vaccine research was focused on the development of a preventive vaccine, which attempted to encourage the immune system to produce **antibodies** that would block the virus from infecting cells. This approach has been quite successful in laboratory experiments; however, to date, all experiments have failed when mutated strains of HIV were introduced to the culture. According to the **AIDS Vaccine Advocacy Coalition** (AVAC), determining how to stimulate an antibody response is "a task that most researchers consider essential for an optimal vaccine but one that's proven impossible so far" (www.avac.org).

The current research efforts to develop a vaccine to combat HIV have shifted to **therapeutic vaccines,** which seem to have a greater potential for success. Therapeutic vaccines are great for targeting chronic infectious diseases, but also noninfectious conditions with massive immune system involvement. Many of the new vaccine candidates that are currently in clinical trials are therapeutic vaccines. A therapeutic HIV vaccine (also known as a treatment vaccine) is a vaccine used in the treatment of an HIV-infected person. Therapeutic HIV vaccines are designed to boost the body's immune response to HIV in order to better control the infection. This type of vaccine is not designed to prevent HIV infection; however, it is hoped that these vaccines will significantly delay the progression to AIDS, thereby greatly improving the quality of life for HIV-infected persons. There are currently no FDA-approved therapeutic

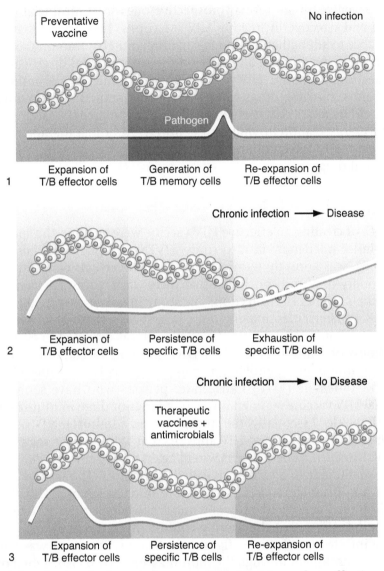

Figure 11-2 (1) Infection is thwarted upon administration of an effective preventative vaccine through induction of T (light blue) and B (light red) effector and memory cells prior to pathogen exposure (white line). (2) Continuous pathogen replication and release of microbial products and exhaust immune responses over the course of a chronic infection leading to a failure to prevent disease. (3) In combination with antimicrobials, **therapeutic vaccines** might restore immune control and prevent disease in chronic infection.

vaccines; however, there are several prospective ones in clinical trial to determine if they are safe and effective in treating HIV infections. It is hoped that if therapeutic vaccines are able to strengthen the body's natural anti-HIV immune response, people with HIV will not have to rely so heavily on the antiretroviral drugs currently used to treat HIV infection. Currently, HIV patients must continue to take antiretroviral drugs for life, and most of these drugs cause serious and sometimes life-threatening side effects. It should be noted that even an effective therapeutic HIV vaccine probably will not be able to replace antiretroviral drugs entirely. At best, a therapeutic HIV vaccine may help control HIV infection and keep people healthy while minimizing the need for antiretroviral drugs.

Failure of the HVTN-505 Vaccine Trial

The challenges of creating an effective HIV vaccine were again in the spotlight recently when a $77-million federally funded study, HVTN-505 (ClinicalTrials.gov Identifier: NCT00865566), was abruptly terminated after the oversight committee took a look at the preliminary data and concluded that there was no way the study would show that the vaccine prevents HIV infection. The study involved over 2,500 participants in 21 sites across 19 large cities, making it the largest current study in the United States. This phase 2, randomized, placebo-controlled trail was designed to evaluate the safety and effectiveness of a multi-clade HIV-1 DNA plasmid vaccine followed by a multi-clade HIV-1 recombinant adenoviral vector vaccine in HIV-uninfected, circumcised men and male-to-female (MTF) transgender persons who have sex with men. The investigational HIV vaccine regimen involved a series of three immunizations over the course of eight weeks. Essentially, the strategy was to administer a DNA-based vaccine made with genetically engineered HIV antigens that would prime the immune system to attack HIV. Later, the same antigens would again be administered to the participant, encased in a shell made of a cold virus called adenovirus. The second administration would act as a booster shot, strengthening the initial response. This the researchers hoped would train T cells to identify and attack the very earliest HIV-infected cells in the body. The hope was that the vaccine could either prevent HIV infection or help those already infected to effectively fight the infection. The study was halted because from the day of enrollment through the month 24 follow-up visit, a total of 41 cases of HIV infection occurred in the study participants who received the vaccine regimen and 30 cases of HIV infection occurred among the placebo vaccine recipients. Moreover, for individuals who were vaccinated and then became infected, the vaccine did not lower their viral load. It was therefore recommended that no further vaccinations with the investigational vaccine regimen be administered. While this outcome was very disappointing for the entire HIV/AIDS research and patient community, there is no doubt that the National Institutes of Allergy and Infectious Disease (NIAID) who sponsored this study as well as the International AIDS Vaccine Institute and their partners, remain committed to finding a highly effective preventive AIDS vaccine.

Logistics of HIV Vaccine Clinical Trials

Successful clinical trials are essential to any new HIV vaccine candidate. These trials allow investigators to administer candidate vaccine and measure their effects under controlled conditions. These trials are divided into three phases, which are designed to test different aspects of safety, efficacy, and overall protection. Phase I trials are usually small, involving 20 to 50 volunteers. They are designed to test a vaccine's general safety. If results are favorable, larger Phase II trials are conducted with approximately 100 to 300 volunteers, in which the vaccine is evaluated to determine if it can stimulate an immune response in humans. In Phase III, the largest of the clinical trial phases, scientists attempt to determine the efficacy of the vaccine: Does the vaccine actually work in preventing the disease? Phase III trials usually involve several thousand healthy volunteers at high risk of HIV infection divided into two groups. Group 1 contains those receiving the experimental vaccine, and group II consists of those receiving a control or placebo treatment. These two groups are monitored for approximately three years, at the end of which the incidence of new infections are compared between the two groups. These results form the basis upon which the efficacy of candidate vaccines is measured. Individuals participating in HIV clinical trials will test positive for HIV antibodies and are therefore only eligible to participate in a single trial.

Once a promising candidate vaccine has been identified and tested in the laboratory and in animal models, it can be moved to clinical evaluation in humans. In order to test the efficacy of a vaccine, large populations with high incidences of the disease—primarily located in developing countries—are required. The study volunteers in these trials are typically individuals who are in high risk groups in various parts of the world. The reason for this is that if volunteers are not exposed to HIV, then you will not know if the vaccine actually works. Therefore, if the results are to show a difference in incidence between the control and the treatment group, new infections must occur among trial participants. The volunteers are not forced or encouraged to participate in unprotected sex. In fact, they are typically counseled on HIV risk factors and how to protect themselves from HIV infection as well as given condoms prior to the start of the study. The reality remains that the effectiveness of any experimental vaccine cannot be fully determined unless it is the only variable being tested and administered in the study.

There is no doubt that tremendous progress has been made in combating HIV infection in the United States and other industrialized countries. However, new infections continue to occur worldwide and there are specific demographic areas that continue to be a major cause for concern. There is, however, an urgent need to find effective measures to prevent new infections, particularly in those parts of the world where HIV continues to spread unabated, and antiretroviral therapies are outside of the financial reach of those in greatest need. Throughout history, vaccines have always been great equalizers in society because they were always made available to both the rich and the very poor—those in developed regions and those in the poverty stricken corners of our planet. While a significant portion of HIV research dollars has been appropriated for vaccine research, most researchers agree that a successful vaccine is no closer than decades away. "Even if we come up with a cure or vaccine tomorrow, just think about the time that would be needed to implement all these measures widely throughout the world," adds Dr. David Ho, director and CEO of the Aaron Diamond AIDS Research Center and the Irene Diamond Professor at the Rockefeller University in New York. "And even with that optimal scenario, it would be decades before this fight is won, and we're certainly, unfortunately, not in that situation" (www.pbs .org/wgbh/pages/frontline/aids/virus/vaccines.html). Development of a safe and effective HIV/AIDS vaccine is an imperative if the global pandemic is to be controlled and ultimately stopped. Clearly, there will always be a need for the important roles played by education, behavioral intervention, and pre-exposure prophylaxis. However, the logistical implications of treating even just the most at-risk HIV uninfected individuals are staggering. Only a safe and efficacious HIV vaccine will be able to provide the type of global control needed to bring an end to the pandemic. Recent evidence suggests that even a partially efficacious vaccine could substantially alter the course of the HIV/AIDS epidemic by lowering infection rates, thus saving lives and at the same time proving cost effective.

Test Your Knowledge

Name: _____

1. True or False: You have to be infected with HIV to participate in a preventive HIV vaccine clinical trial.

2. True or False: Before people volunteer for preventive HIV vaccine clinical trials, they get detailed information on the side effects they might experience.

3. True or False: There are currently no preventive HIV vaccines available for human use.

4. True or False: Subunit vaccines contain only individual parts of HIV, rather than the whole virus.

5. The only way an HIV vaccine study volunteer could get infected with HIV is if they_____.

   (A) received the HIV vaccine instead of the placebo
   (B) hugged someone infected with HIV
   (C) didn't use a condom with someone infected with HIV
   (D) had a cold at the time they participated in the study

6. A vaccine composed of attenuated viruses _____.
   (A) has viruses that multiply in the body
   (B) contain viruses that have been chemically inactivated
   (C) cannot be used to induce immunity
   (D) is in the developmental stage for AIDS

7. All of the following may surface as valid concerns during an HIV vaccine trial except_____.
   (A) volunteers can only participate in a single trial
   (B) counseling must be given to all recipients of the vaccine
   (C) volunteers will have to contend with replicating viruses from the vaccine
   (D) participating in a vaccine trial will result in a positive test of HIV antibodies

135

8. Integrase inhibitors work by _____.

   (A) binding to reverse transcriptase
   (B) blocking CD4 receptor and preventing gp120 from attaching
   (C) preventing HIV nucleic acid from being spliced into host DNA
   (D) blocking gp41

Answers to Review Questions

Chapter 1

1. True
2. True
3. False
4. (E) Both A and D
5. (C) Kaposi's sarcoma
6. (C) 69%
7. 34 million
8. 1981

Chapter 2

1. (B) It contains either DNA or RNA
2. (E) It cuts viral polyprotein into individual functional proteins
3. (B) The HIV organism lives inside cells and body fluids
4. (C) is from the lentivirus family
5. Reverse transcriptase
6. True
7. Dmitri Iwanowski was the first person who discovered viruses

8. The *gag* gene (group-specific antigen) codes for the precursor gag polyprotein which is processed by viral protease during maturation to matrix protein (p17) and capsid protein p24). The *pol* gene codes for codes for viral enzymes reverse transcriptase and RNase H, integrase, and HIV protease. The *env* gene codes for structural envelope glycoproteins gp160 which is cleaved into gp120 and gp41.

Chapter 3

1. (B) simian immunodeficiency virus (SIV) in chimpanzees
2. False (HIV traveled from the Congo, to Haiti, and then to the United States)
3. True
4. Peter Duesberg
5. HIV came about through viral sex where two different viruses entered the same cell and through exchange of genetic material a third virus was formed, which later became HIV, possibly after multiple additional mutations.
6. The strongest evidence that HIV causes AIDS is that the higher the viral load, the greater the probability of developing AIDS. Also, antiretroviral drugs are able to reduce the viral load and improve the health of the HIV-positive patient.
7. (B) Subtype B is the most common HIV strain in North America

Chapter 4

1. (D) Activate the cell-mediated and antibody-mediated immune defenses
2. (B) CXCR4 tropic, when Gp120 on infected cell bind CD4 on the surface of uninfected cells
3. (B) Gp120, CD4
4. (C) Plasma cells
5. False. The innate and adaptive immune responses complement each other in an attempt to remove an invader.
6. True
7. CXCR4-tropic viruses are associated with more pronounced depletion of CD4 T cells because most T cells have CXCR4 receptors. Therefore, CXCR4 viruses have a wider range of susceptible target cells, leading to higher viral load and lower CD4 T cell counts.
8. The normal function of the CD4 receptor on a T cell is to assist in a process called antigen presentation where professional antigen presenting cells process and present peptides from invaders to helper T-cells leading to their activation and thereby eliciting the T-cell help to respond to the invading microbe.

Chapter 5

1. (E) Both C and D
2. (A) AIDS dimentia complex
3. (C) *Pneumocystis carinii* pneumonia
4. (C) when his or her immune system is seriously compromised
5. (A) CD4 cell count below 200 and high viral load
6. The window period of HIV infection is the time from infection to the first appearance of detectable HIV antibodies in the patient's serum. During this time HIV antibody tests will be negative although the viral RNA will be detectable.
7. Seroconversion is the appearance of detectable antibodies to HIV in the patient's serum.

Chapter 6

1. (E) Both C and D
2. (B) STDs can cause erosion of the urethra and provide an entry path for HIV
3. (B) social kissing
4. (D) All of the above
5. True; This is a part of the universal precaution guidelines

Chapter 7

1. (D) sub-Saharan Africa
2. (C) Unprotected heterosexual sex
3. (A) Men who have sex with men
4. True
5. China has the "Four Frees and One Care" as well as the "Five Expands, Six Strengthens" programs to combat HIV/AIDS

Chapter 8

1. (B) Blacks
2. (B) 1%
3. True
4. True
5. False. The southern U.S. states have the greatest HIV prevalence rate

6. True

7. True

8. The U.S. city with the highest HIV/AIDS prevalence rate is Washington, DC

Chapter 9

1. False. There are currently no FDA-approved HIV testing kits that can be administered and interpreted by the patient.

2. True

3. The U.S. Food and Drug Administration

4. (B) ELISA

5. (C) Four weeks

6. (B) Within 30 minutes

7. (B) HIV antigen and antibodies

8. Newborns do not have a developed immune system and would therefore not have HIV-specific antibodies in their blood. Moreover, newborns receive antibodies from their mothers through the placenta. The HIV antibody test could recognize maternal antibodies rather than those of the child who is exposed to HIV. You therefore have to test for the presence of the virus directly.

9. Genotyping tests are useful in identifying mutations associated with resistance to reverse transcriptase and protease inhibitor. Phenotyping helps predict viral load response to new antiretroviral drugs in individual patients.

10. CCR5 is a co-receptor on the surface of CD4+ helper T cells that are required for HIV binding and entry. Entry inhibitors block CCR5, preventing the viral particle from binding to it. HIV therefore cannot enter the cell to cause a productive infection.

Chapter 10

1. (D) NNRTIs

2. (B) Atripla

3. (A) binding to Gp41 and preventing HIV entry

4. (B) The amount of virus in the blood

5. (A) The donor had a homozygous CCR5 mutation

6. True

7. True

8. False (there should be at least three drugs at any one time)

9. The Mississippi baby apparently had aggressive early drug therapy that killed off the HIV before it could establish a hidden reservoir in the lymph nodes, spleen, and bone marrow.

10. Melittin. The mechanism of action appears to be its ability to poke holes in the viral envelope.

Chapter 11

1. False; preventive HIV vaccine clinical trials do not enroll seropositive individuals
2. True
3. True
4. True
5. (C) didn't use a condom with someone infected with HIV
6. (A) has viruses that multiply in the body
7. (C) Volunteers will have to contend with replicating viruses from the vaccine
8. (C) Preventing HIV nucleic acid from being spliced into host DNA

Selected HIV/AIDS Online Resources

http://kff.org/hivaids/: **The Kaiser Family Foundation** is a non-profit foundation that focuses on major health care issues affecting global health policies and seeks to highlight US involvement. The foundation funds its own independent research and has strong collaborations with several major non-profit organizations and media groups. The site provides up-to-date statistics and relevant policy information on HIV/AIDS and several other relevant diseases.

http://www.unfpa.org/aids_clock/index.html#: **The AIDS Clock** was created in 1997 and became web-based in 2000. The United Nations Population Fund created the clock as a way to acknowledge both the toll of the epidemic, and the partnership that was formed to tackle it, **UNAIDS** (the United Nations Joint Program on HIV/AIDS). While the precise numbers of people living with HIV, people who have been newly infected, or who have died of AIDS are not known, the **AIDS clock** provides point estimates in real time along with other useful statistics on AIDS.

http://aids.gov/: This site is a one-stop shop for HIV-related Federal policies as well as programs and resources on basic HIV/AIDS information. With is integrated social networking platform, this site makes it easy to communicate with and get information from several sources including Facebook and Twitter.

http://www.aidsinfo.nih.gov/: **AIDS***info* is a U.S. Department of Health and Human Services (DHHS) website that offers the latest federally-approved information on

HIV/AIDS clinical research, treatment and prevention, and medical practice guidelines for people living with HIV/AIDS, their families and friends, health care providers, scientists, and researchers.

http://www.amfar.org/: **The American Foundation for AIDS Research (amfAR)** is the nation's leading nonprofit organization dedicated to the support of HIV/AIDS research. The organization seeks to develop prevention methods (including a vaccine), improve treatments, and ultimately find a cure for AIDS. amfAR-funded research makes significant contributions to the lives of people with HIV/AIDS and to the global effort to arrest the epidemic.

http://www.avert.org/: **AVERT,** an international HIV and AIDS charity based in the United Kingdom, has a number of overseas projects, helping with the problem of HIV/AIDS in countries where there is a particularly high rate of infection, such as South Africa, or where there is a rapidly increasing rate of infection, such as in India. This site is very comprehensive with up-to-date statistics and touches on all of the important issues.

http://www.hivinsite.org/: **HIV InSite** is a source for comprehensive, in-depth HIV/AIDS information and knowledge developed by the **Center for HIV Information (CHI) at the University of California San Francisco (UCSF)**, http://chi.ucsf.edu/.

http://www.nytimes.com/library/national/science/aids/aids-index.html: **The** *New York Times* hosts this wonderful site containing current relevant news about HIV/AIDS as well as audio and video multimedia resources organized by dates in various categories. There is also a great archive of articles published by the *New York Times* over the years.

http://www.thebody.com/index.html: **The Body** is a comprehensive site with research-based HIV/AIDS resources, including treatment and prevention, as well as forums where you can ask relevant questions to experts in the field. The Body also hosts the "Visual AIDS Web Gallery" with art created by HIV-positive artists.

http://www.cdc.gov/hiv/dhap.htm: **The Center for Disease Control and Prevention (CDC)** is the major United States government-run resource for facts about HIV/AIDS in the United States. The site includes information on prevention programs, fact sheets about different high-risk groups, and statistics broken down by state. This is the site on which the first cases of HIV infection were published and it continues to host a collection of articles from the CDC's Morbidity and Mortality Weekly Report on HIV/AIDS.

http://www.iavi.org/: **The International AIDS Vaccine Initiative (IAVI)** is a global not-for-profit organization working to speed the search for a vaccine to prevent HIV infection and AIDS. Founded in 1996 and operational in 23 countries, IAVI and its network of partners research and develop vaccine candidates. IAVI also advocates for a vaccine to be a global priority and works to assure that a future vaccine will be accessible to all who need it.

http://www.unaids.org/: **UNAIDS, the Joint United Nations Program on HIV/AIDS,** brings together the efforts and resources of 10 UN system organizations to the global AIDS response. The site brings to the forefront the many challenges facing individuals with AIDS as well as the progress being made in education, treatment, and new initiatives.

http://www.worldaidscampaign.info/: **The World AIDS Campaign (WAC)** is an organization that has been established to strengthen and connect together the advocacy and campaign activities targeting governments and other AIDS organizations. The vision of the WAC is to establish a global movement bringing renewed impetus and resolve to the fight against the epidemic. The WAC is currently responsible for organizing the annual **World AIDS Day** on December 1st each year.

Index

Note: Figures are indicated by an italic 'f'; tables are indicated by an italic 't'.